Contents

GEOLOGY
along
SKYLINE DRIVE

Shenandoah National Park,
Virginia

Robert L. Badger

You can order extra copies of this book and get information and prices for other Falcon guidebooks by writing Falcon, P.O. Box 1718, Helena, MT 59624 or calling toll-free 1-800-582-2665. Please ask for a free copy of our current catalog. Visit our website at http://www.falconoutdoors.com

© 1999 Falcon® Publishing, Inc., Helena, Montana.
Printed in Korea.
All black-and-white illustrations © 1998 by Sue Thomas.

1 2 3 4 5 6 7 8 9 0 CE 04 03 02 01 00 99

Falcon is a registered trademark of Falcon® Publishing, Inc.
All rights reserved, including the right to reproduce this book or any part thereof in any form, except for inclusion of brief quotations in a review.

Inside photos and cover photo by Robert Badger.

Cataloging-in-Publication Data is on record at the Library of Congress.

CAUTION
All participants in the recreational activities suggested by this book must assume responsibility for their own actions and safety. The information contained in this guidebook cannot replace sound judgment and good decision-making skills, which help reduce risk exposure; nor does the scope of this book allow for disclosure of all the potential hazards and risks involved in such activities.

Learn as much as possible about the recreational activities in which you participate, prepare for the unexpected, and be cautious. The reward will be a safer and more enjoyable experience.

Foreword

The longer I spend in and around Shenandoah National Park, the more I am convinced that geology is the driving force of the human and natural system. Differences in land use and human history can be explained, at least in part, by factors that derive from the geologic history and underlying bedrock in different areas: the terrain, soil development and fertility, water availability, and even vegetation are easier to understand when you have a basic understanding of the bedrock.

While this is true almost anywhere, at Shenandoah it has particular meaning for two important reasons. First, the susceptibility of the park's soils, waters, and biota to the devastating effects of air pollution and acid rain is directly tied to the nature and chemistry of the three major rock types in the park. And second, the major landslides and debris flows in and near the park in 1995 were a lesson to all who were here that what is on the surface of the Blue Ridge Mountains may be what we are most familiar with, but it is but a thin skin on a range of ancient mountains that is still very geologically active.

Because of the park's thick mantle of vegetation, it is not easy for most visitors to the park to see, much less understand, the underlying geology and its significance. A knowledgeable guide certainly helps and can make the geologic story fascinating as well. This is where Rob Badger comes in.

Dr. Badger has been working in Shenandoah with National Park Service research permits, on and off, since 1984. In the course of his research, he has developed a unique understanding of the Catoctin metabasalts, one of the principal rock types in the park. Although he lives hundreds of miles away in northern New York, he comes back time and again as a visitor and has retained the sense of fascination and wonder that most of us felt the first time we traveled through this special place we call Shenandoah National Park. I was quite pleased when he first broached the subject of writing a guidebook to the accessible geology of the park.

I know I will keep a copy of *Geology Along Skyline Drive* in my car and expect it to be dog-eared before too long. I hope you will, too. ■

Bob Krumenaker

CHIEF, DIVISION OF NATURAL AND CULTURAL RESOURCES
SHENANDOAH NATIONAL PARK
NOVEMBER 13, 1997

Preface

Geologists have sometimes been accused of finding beautiful places to work and then finding a project to work on there. To that charge, I plead guilty. When I started my graduate studies at Virginia Polytechnic Institute in the fall of 1983, my wife was pregnant, so I could not go gallivanting off to Alaska, California, or Maine the following summer, as some of my fellow students did. Other colleagues who chose their research areas close to home were working in the Virginia and North Carolina Piedmont, where it was hot, humid, and buggy. I chose the high ground of the cool Blue Ridge Mountains where the marvelous Catoctin metabasalts cover much of Shenandoah National Park. To these rocks and peaks I return time and again, and I always feel at home when I am hiking among them. ■

Acknowledgments

I am indebted to Krishna Sinha, who first introduced me to the ancient Catoctin volcanics and the multitude of problems that they posed for a geologist, and who remained my mentor throughout my studies at Virginia Tech.

Previous work on the Catoctin Formation in Shenandoah National Park by Jack Reed during the 1950s, and work on the geology of the entire park by Tom Gathright during the 1960s and 1970s, greatly facilitated my own studies; without the efforts of these geologists, my scholarly research and this manuscript would have been much more difficult.

I am grateful to the State University of New York at Potsdam for granting me sabbatical to work on this manuscript, providing time to spend a wonderful three months during the fall of 1997 to hike the trails, take pictures, and examine the geology of Shenandoah National Park.

Scientists at the University of Virginia, especially Rick Webb, Tanya Furman, and their colleagues in the Department of Environmental Sciences, have been most helpful in aiding my understanding of the acid precipitation and the ability of the bedrock to neutralize these acids in Shenandoah National Park.

Many members of the National Park Service, including Tom Blount, Karen Michaud, and Bob Krumenaker, along with Greta Miller of the Shenandoah Natural History Association, are to be thanked for their assistance and encouragement. My friend Sue Thomas contributed the wildlife sketches for this project, and she, along with Bill Jaggers, did the technical freehand drawings.

Rick Webb constructed the simplified geologic map of the park and the chart comparing sulfate in precipitation at various national parks. Jeff Perkins, Chris Cleveland, and Bill Wiese assisted me with the other computer-generated diagrams. Reviews by Jim Carl, Ben Morgan, John and Sue Omohundro, Ken Beyer, Joanne Amberson, and Rock Comstock greatly improved the manuscript. ■

About This Tour

Shenandoah National Park offers its visitors a multitude of opportunities to observe nature. The wildlife is spectacular, the botany is terrific, and the geology is intriguing. As with every subject, the more you understand geology, the more you can appreciate it. Such features as columnar jointing, amygdules, volcanic breccias, and Skolithos tubes tell us volumes about the geologic history of this park, but to most visitors these features remain unrecognized.

I have seen at least a dozen people take pictures of their friends perched on top of a huge boulder of volcanic breccia at Franklin Cliffs Overlook, but I have yet to see one person take a look at the rock, a rock which I find to be fascinating. This book was written for people who are interested in learning more about the geology of the park, about the different geologic features that can be observed here, and who would like to know exactly where to go in the park to see some of these features.

This book is a guide to the geology along, and adjacent to, Skyline Drive in Shenandoah National Park. Within the park, Skyline Drive stretches for 105 miles along the crest of the Blue Ridge Mountains. Along the drive, there are approximately 75 scenic overlooks and numerous parking areas at trailheads. Interesting geologic features are readily accessible and can be observed at almost all of these overlooks and trailheads, and, if one had the time, a week or more could easily be spent examining them all. But few have the luxury of such free time, so for the purposes of this book, I have selected only the most interesting and accessible locations along Skyline Drive from which to observe the geology.

A stop at every location in this guide, followed by the recommended short hikes at many of the stops, would be nearly impossible to do in a single day. Just driving from one end of Skyline Drive to the other at the posted 35 mph speed limit takes more than three hours, but add time for the sharp curves, morning fog, slow moving vehicles, and indifferent deer that wander the road at will. So, if you have but one day, pick and choose, perhaps staying in one section of the park (see overview map on page x) and visiting the recommended geological features at a slow and steady pace. Take the hikes, as well; most are on well-groomed trails with solid footing. Plan on spending two days on the tour, perhaps with a night at Skyland or Big Meadows Lodge (reservations recommended; call 1-800-999-4714) or at one of Shenandoah's four campgrounds located along Skyline Drive.

The stops are located either at overlooks, which are well marked by signs along the drive, or at parking areas at trailheads, which have smaller and more discreet signs. Mileages in this guide are based on the cement mileposts that are along the entire length of Skyline Drive, beginning with mile 0 at the north entrance to the park and ending at mile 105 at the south entrance, so you need not set an odometer or worry about calculating the mileage, even if you are traveling from south to north.

Because I expect readers to pick and choose among the stops or to be traveling in reverse or random order, there is some duplication in the explanations of the various rock features. Where a feature is not explained, quickly look to the descriptions of similar or adjacent stops, and I am sure you will find an informative discussion. I have tried to stay away from technical jargon, but I have had to use a number of basic geologic terms. Where these terms first appear in the text, they are *italicized* and can be found in the glossary at the end of the book.

Please remember that this is a national park. Collecting specimens of rocks or plants, or using rock hammers, is prohibited. ■

Index Map of Shenandoah National Park

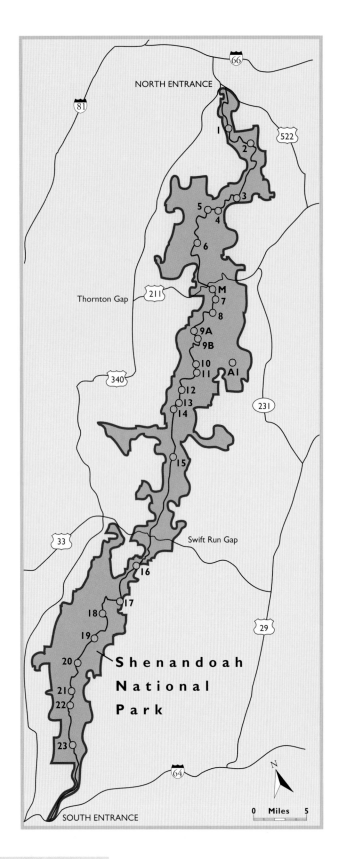

1. Signal Knob Overlook
2. Indian Run Overlook
3. Range View Overlook
4. Little Devils Stairs Overlook
5. Hogback Overlook
6. Thornton River Trail Parking Area
M. Marys Rock Tunnel
7. Hazel Mountain Overlook
8. Pinnacles Overlook
9A. Stony Man Overlook
9B. Little Stony Man Parking Area
10. Crescent Rock Overlook
11. Hawksbill Gap Parking Area
12. Franklin Cliffs Overlook
13. Dark Hollow Falls Parking Area
14. Big Meadows Area, Blackrock
15. Bearfence Mountain Parking Area
16. Bacon Hollow Overlook
17. Loft Mountain Overlook
18. Rockytop Overlook
19. Doyles River Overlook
20. Blackrock Parking Area
21. Horsehead Mountain Overlook
22. Riprap Parking Area
23. Sawmill Ridge Overlook
A1. Old Rag Mountain

General Geology of Shenandoah National Park

Skyline Drive offers a 105-mile-long opportunity to observe the geology of Shenandoah National Park in the Blue Ridge Mountains of northern Virginia. Initiated by President Herbert Hoover in 1931, and expanded in 1933 by President Franklin Roosevelt as a Bureau of Public Roads project, the road runs along the ridgeline and has marvelous views from numerous scenic overlooks. The Appalachian Trail also runs the length of the park, roughly parallel to Skyline Drive and crossing it in numerous places. There are many side trails, loop trails, and trails to waterfalls and mountaintops that are generally short and gentle since the road traverses the ridgeline.

Traveling along Skyline Drive, whether by foot, automobile, bicycle, or motorcycle, you cannot help but notice the cliffs and ledges of rocks rising from the roadside, many of which were exposed when Skyline Drive was constructed. These steep-walled slices through the hills expose what is called the *bedrock*, the ancient, solid rock that composes the mountains. Here are exposed examples of all three rock groups—*igneous, metamorphic,* and *sedimentary*—as well as interesting features in the rocks that help to unravel the geologic history of the park. There is much to learn from roadcuts if one takes the time to study them.

SHENANDOAH'S THREE BASIC ROCK TYPES

Look closely at the rocks. One type has large white, tan, and grayish mineral grains with a few scattered black crystals. Another is dark green, fine grained, and very hard. A third kind resembles grains of sand that have been glued together to make rocks in a variety of colors—tan, gray, dark red, and greenish gray. These represent the three basic varieties of rock in the park: *granites, basalts,* and *siliciclastic* rocks, which are sedimentary rocks that contain abundant silica or sand (see map on page 2 and columnar section in figure 1).

Granites

The white-to-tan rocks with scattered dark minerals are granites. These are coarse-grained, crystalline rocks containing *quartz, feldspar,* and small amounts of dark, iron-bearing minerals, usually *pyroxene, hornblende,* and *biotite,* a black mica. The grains have interlocking crystal faces and are large enough to be seen with the unaided eye. Some feldspar crystals are more than an inch in diameter. The quartz is

1

Shenandoah National Park
Major Bedrock Types

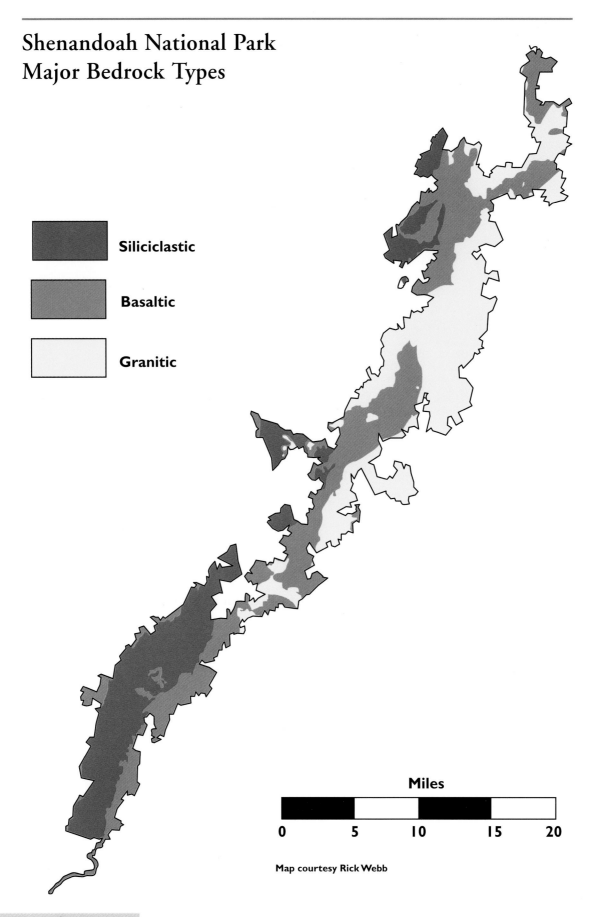

Siliciclastic

Basaltic

Granitic

Miles

| 0 | 5 | 10 | 15 | 20 |

Map courtesy Rick Webb

translucent and gray to blue-gray in color. There are two types of feldspar: *ortho-clase feldspar*, which is often milky colored and sometimes pinkish, and *plagioclase feldspar*, which is usually white or cream colored.

Rocks composed of these minerals include granite, *granodiorite*, which differs from granite only in that it contains more plagioclase than orthoclase, and granitic *gneiss*, a metamorphic rock that is compositionally similar to granite, but the recrystallized minerals are aligned in layers, giving the rock a banded appearance. Along Skyline Drive, these rocks are visible between mileposts 20 and 21 (Stop 5), between Thornton Gap and Little Stony Man (Stop 7), just south of Swift Run Gap (Stop 16), and at Old Rag Mountain (Appendix 1). These rocks are referred to as the Pedlar Formation and, on Old Rag Mountain, as the Old Rag Granite.

Basalts

The dark green rocks that form most of the roadside *outcrops* from the north entrance of the park, 31.4 miles south to Thornton Gap, and for another 26 miles from Stony Man to Swift Run Gap, are perhaps the most noticeable to the park visitor. Stop at Crescent Rock or at Franklin Cliffs and look at the view, then look beneath your feet. There you will find the subject of my studies as a geologist and student at Virginia Polytechnic Institute, the old green volcanic rocks of the Catoctin Formation.

These green rocks were once basalts, a common type of volcanic rock containing the minerals pyroxene and plagioclase feldspar. The basaltic lava of the Hawaiian Islands is composed of this same type of rock. However, there are some key differences. Basalts (like the ones in Hawaii) are usually black, but the rocks in Shenandoah National Park are dark green, and a close look seldom reveals the minerals pyroxene and plagioclase feldspar. What happened to Shenandoah's basalts to make them appear dark green and change their mineralogy? The answer is that they were altered by a process known as *metamorphism*, which involves alteration of the rocks by heat and pressure while slowly percolating hot fluids (mainly water) through them.

This process occurred in Shenandoah when the Appalachian Mountains were formed and metamorphism caused the old minerals to recrystallize into new ones. The new minerals, visible with a hand lens or microscope, are *chlorite*, a green mica which gives the rock its green color; *epidote*, a pistachio- to olive-green mineral; and *albite*, a type of feldspar that contains sodium (as opposed to plagioclase feldspar, which contains calcium, or orthoclase, which contains potassium). These altered basalts can no longer be called basalts, but instead are called *metabasalts* or, sometimes, *greenstones*. They have been assigned to a grouping of lavas and

Figure 1: Stratigraphic Units

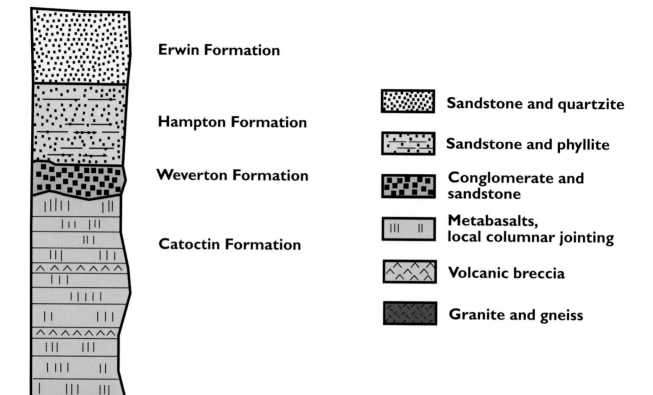

Erwin Formation

Hampton Formation

Weverton Formation

Catoctin Formation

Swift Run Formation

Pedlar and Old Rag formations

Sandstone and quartzite

Sandstone and phyllite

Conglomerate and sandstone

Metabasalts, local columnar jointing

Volcanic breccia

Granite and gneiss

sedimentary rocks that collectively make up the Catoctin Formation. The term *formation* refers to a mappable unit of rock that can be distinguished by its rock type from surrounding rock units.

Sedimentary Rocks

The third type of rock consists of sedimentary rocks, including *sandstone, conglomerate, phyllite,* and *quartzite.* They are found mostly in the southern part of the park and in a small area north of Thornton Gap (see Stop 6). Sandstones are composed of quartz grains that have been compressed to form rock. They are usually tan to gray, but some are greenish gray or dark red, indicating the presence of green chlorite and epidote or a red iron oxide mineral called *hematite.* Conglomerates contain coarser material, such as pebbles that can be found in a streambed. Phyllites are finer grained rocks than sandstone, originally containing clay or mud that was altered to mica when the rocks were metamorphosed. They are thinly layered and often crumble easily. Quartzites are very pure sandstones in which the quartz grains are tightly cemented.

Since all of these rocks contain abundant silica, and all are composed of fragments of older rocks that have been cemented to form new rock, they can be called siliciclastic. Siliciclastic rocks were deposited by streams, rivers, lakes, or even the ocean. These rocks have been assigned to the Weverton, Hampton, and Erwin formations where they overlie the Catoctin Formation. Conglomerates, sandstones, and phyllites below the Catoctin volcanics belong to the Swift Run Formation.

AGE AND ORIGIN OF THE ROCKS

The oldest rocks in Shenandoah National Park are the granites and gneisses, which have *radiometric* dates between 1 and 1.2 billion years old (see Appendix 3 for a discussion of age determination). These rocks underlie the other rock types and are commonly called the "basement" rocks. Most of them crystallized grain by grain from a *silicate melt*, the magma that filled the chambers below ancient volcanoes. As the magma cooled, it crystallized to form the large, interlocking crystals that we observe in the Pedlar Formation and Old Rag Granite. Included in this group of rocks are metamorphic gneisses that were once sedimentary rocks, before they were intensely altered at a later time by heat and pressure.

Figure 2

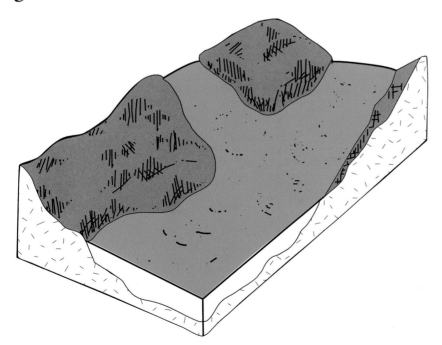

Sketch of magma flowing down preexisting channels during early phases of Catoctin magmatism. Successive flows eventually covered most or all of the land.

Figure 3

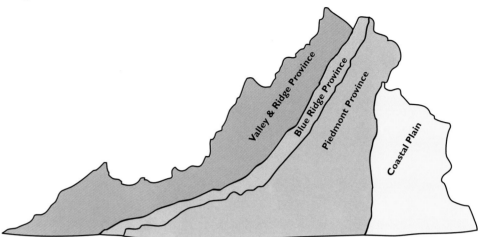

These granites and associated gneisses were part of a mountain range, long since eroded, called the *Grenville Mountains,* which once extended from Texas to Newfoundland and were perhaps as high as the Himalayas of today. All that is left of these mountains are their roots: the granites, granodiorites, and granitic gneisses that are observed in Shenandoah National Park.

Overlying the basement rocks are the metabasalts of the Catoctin Formation, which have been radiometrically dated at approximately 570 million years old (see Appendix 3 for a discussion of age determination). They represent episodic outpourings of basaltic magma that flowed over the land. This magma originated deep within the Earth's mantle and rose to the surface along vertical or near vertical fractures in the Earth's crust. When each volcanic episode ended, magma remaining in the fractures crystallized to form thin, nearly vertical bodies of basalt called *dikes.*

Look for dikes at Little Devils Stairs Overlook (Stop 4), at Marys Rock Tunnel (not a stop because there is no place to park, but you can see the feature from the road as you drive by), and at Old Rag Mountain (Appendix 1).

As this tour demonstrates, the lava flowed out in pulses. The first pulse poured into the river valleys and left the higher ground exposed (see figure 2, page 5). Successive lava flows covered the higher elevations until the old landscape was completely submerged. Individual lava flows were generally 30 to 90 feet thick near the source of the dikes that fed them. A single outpouring of lava might have had a duration of several weeks or even months, as the volumes of extruded magma were enormous. Eventually, the lava slowed and stopped as the magma crystallized to rock. A period of quiescence would follow for hundreds or perhaps thousands of years before another pulse of magma surfaced and flowed over the land. Between eruptions, streams and rivers formed, leaving thin deposits of gravel, sand, and mud on top of the newly crystallized lava flows. When the next volcanic event occurred, the lava either flowed over the sedimentary layers or simply plowed into them, scooping up and incorporating the material into the moving magma. In Shenandoah National Park, we see evidence of both occurrences.

The sandstones, conglomerates, phyllites, and quartzites are sedimentary rocks that are grouped by geologists into four different rock formations in the region. One of these, the Swift Run Formation, was deposited by streams and rivers on top of the rocks of the Pedlar Formation before eruption of the Catoctin volcanic rocks (Stops 11 and 15). This rock unit is not a continuous layer between the Pedlar and Catoctin formations, but is only found at what are generally interpreted to have been old river valleys and flood plains underneath the Catoctin Formation. The other three formations were deposited on top of the volcanic rocks and are therefore younger. The lowermost of these three rock units, called the Weverton Formation (Stops 6 and 23), consists of sandstones and conglomerates that were deposited by rivers and streams. Overlying this is the Hampton Formation (Stops 6, 19, and 20), which consists mostly of tan-to-gray sandstone and quartzite. Rocks of the Hampton Formation were primarily beach sands deposited along the shores of a shallow inland sea. Some thin beds of phyllite dispersed between the sandstone beds can be seen at Stop 19. The phyllites represent sediments of mud or clay that were deposited in lagoons. The youngest formation to be seen on this tour is the Erwin Formation, a white quartzite that overlies the Hampton Formation. The Erwin Formation was also beach sand. It contains fossils of *Cambrian* age, which helped to determine the age of the Catoctin basalts that lay below (see Stop 22 and the discussion on age determination in Appendix 3).

PHYSIOGRAPHIC PROVINCES

Three physiographic provinces occur in central and western Virginia: the Blue Ridge, the Valley and Ridge, and the Piedmont (see figure 3). A *physiographic province* is a region of similar geologic features and landforms that experienced the same geologic history, whereas adjacent regions have geologic features and landforms that are quite different.

The Blue Ridge Province

The Blue Ridge, of course, is the setting for Shenandoah National Park. It is part of the Appalachian Mountains and consists of one main ridge of that chain. A few mountains and hills not directly linked to the main line are assigned to the Blue Ridge as well, but these are smaller features and usually stand alone. The highest point along the Blue Ridge in Shenandoah National Park is 4,049 feet above sea level, at the top of Hawksbill Mountain. Elevations decline rapidly as you descend from the ridgeline to the valleys to the east and west.

The Valley and Ridge Province

West of the Blue Ridge Mountains is the Shenandoah Valley, bisected by Massanutten Mountain, a long, linear ridge. West of the valley is another ridge, and then another, each flanked by elongate valleys. This is the Valley and Ridge Province, which is characterized by long, parallel, even-topped, generally hard sandstone ridges separated by long, low-lying valleys composed of softer rocks that are more prone to erosion, such as *limestone* and *shale*. These rocks are younger than the sedimentary rocks found in Shenandoah National Park, but they were deposited in the same ocean basin as Shenandoah's strata. Many of the ridges have low-angle faults, called *thrust faults*, bounding their west side (see figure 7, page 21). The thrust faults were formed during a time of continental collision which occurred east of the park area. The forces from the continental collision caused land masses in the Valley and Ridge to be pushed to the west along these faults.

The Piedmont Province

East of the Blue Ridge is the Piedmont Province, a gently undulating plateau with numerous small hills and ridges (see photo 7-3, page 31). These hills do not form continuous ridges; they appear as mounds dotting the landscape like oblong spots on a leopard. Much of the province has a deep, reddish soil, indicative of rock weathering that has taken place over many millions of years. Bedrock in the Piedmont consists of ancient metamorphic rocks, the youngest of which date from about the same time as the Catoctin metabasalts.

The Piedmont area has a very complex and poorly understood geologic history. Current theories suggest that some bedrock in the Piedmont is not indigenous to North America but may have originally been part of another continent that collided with North America. Much later, when this other continent split from North America, part of it was left behind. Furthermore, some parts of the Piedmont are believed to have been island arcs, much like Japan or the Philippines, which collided with North America and became part of the continent. Whatever their source, these extraneous fragments of land are now part of the gently rolling Piedmont Province.

PLATE TECTONICS

Plate tectonics is the theory that the Earth's crust is divided into several large plates, some containing continents, some containing crust of the ocean floor, and some containing both. As the plates move, friction along their margins causes earthquakes. If two plates move toward one another, the resulting collision causes volcanoes and mountains to be formed. The Himalayas, Andes, and Appalachians are examples of mountain chains that were formed by the collision of different plates.

Conversely, when two plates pull away from one another, a rift zone is formed; this action causes the crust, either continental or oceanic, to thin, forming a valley. As the crust thins, basaltic magmas, derived from the partial melting of rocks in the Earth's mantle, may rise to the surface through fractures and flow over the land. These basalts are called *flood basalts* since they tend to inundate an entire area. As the plates move apart, the valley expands and the ground level slowly subsides or settles to a lower elevation. If the valley sinks below sea level, water gradually flows into it, eventually forming an ocean basin. This process, from the initiation of a rift zone to the eventual opening of an ocean basin, takes several million years. Both the Atlantic and Pacific oceans were formed by rift zones, and today the East African Rift Zone is splitting the continent of Africa into two parts. Typically, the plates move from 1 to 3 inches per year.

Plate tectonics played a major role in the formation of Shenandoah National Park. During the last 1.2 billion years, four collisional events involving eastern North America have been largely responsible for the formation of mountains and metamorphism of the rocks. The first collisional event, called the *Grenville Orogeny*, occurred about 1 billion years ago and formed the magmas that crystallized to become the granites and granodiorites of the Pedlar and Old Rag formations. The other three collisional events, at roughly 450, 350, and 300 million years ago, resulted in the metamorphism of the rocks and the uplift that formed the Blue Ridge Mountains.

In addition to these collisional events, there have been two rifting events which resulted in the eruption of volcanic rocks and the opening of ocean basins. The first rifting event resulted in the extrusion of the Catoctin volcanic rocks, about 570 million years ago, and the subsequent deposition of the overlying sedimentary rocks of the Weverton, Hampton, and Erwin formations. The second rifting event, approximately 200 million years ago, resulted in the opening of the Atlantic Ocean basin. This information, when placed in chronological order, helps construct the geologic history of the Shenandoah area.

Figure 4

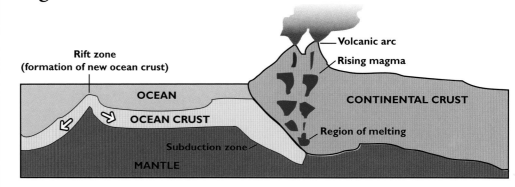

Rift zone
(formation of new ocean crust)

Volcanic arc

Rising magma

OCEAN

OCEAN CRUST

CONTINENTAL CRUST

Region of melting

Subduction zone

MANTLE

Cross-sectional sketch of plate boundaries showing rift zones, where two plates are splitting apart, and collisional zones where two plates are colliding. In the collisional zone, one plate is pushed, or subducted, beneath the other, and it begins to melt at a certain depth, resulting in the formation of volcanoes above.

BRIEF GEOLOGIC HISTORY

Geology is the study of the physical nature and evolution of the earth. Geologists study, among other things, the various minerals, rock types, structures, and mountain belts of a region and try to determine how each was formed. Putting the events that formed these geologic features into sequential order provides a geologic history of the region and allows us to understand the changes that have taken place in that region over a time span of tens or hundreds of millions of years.

Pick a spot, and the geology there may reveal the presence of a barren landscape hundreds of millions of years ago. Perhaps a hundred million years later this spot may have been located in the center of a large volcanic field; another hundred million years later it may have been under an ocean. By understanding the rock types, how they were deposited, and the plate-tectonic setting in which they were formed, we can construct the geologic history of the area now known as Shenandoah National Park. Figure 5 shows the geologic time scale and the timing of geologic events that created the landscape in and around the park.

The oldest rocks, the granites and gneisses of the Pedlar and Old Rag formations, were once part of the Grenville Mountains. They formed during a continental collision about 1 billion years ago and stretched from Newfoundland to Texas. Some geologists believe they may have continued into Mexico. The event that formed this mountain chain, the Grenville Orogeny, also caused the metamorphism of the rocks that existed at that time in Shenandoah National Park.

By 600 million years ago, the Grenville Mountains were gone, eroded to low, rolling hills, and most of their material was washed into the sea. Plants had not yet evolved on land, so these hills would have been devoid of vegetation. The rivers and streams, through their meanderings, left deposits of sand, gravel, and mud; in some places the deposits were nearly 200 feet thick, while other places remained free of sediment. These deposits of varying thickness are now assigned to the Swift

Figure 5: Geologic Time Scale

Geologic Time Period	Time before present (Millions of years before present)	Events in Shenandoah National Park
Quaternary	0 — 2	Erosion
Tertiary		
Cretaceous	65 — 140	Dinosaurs may have roamed in Virginia. Fossilized footprints have been found in Culpeper, east of the park.
Jurassic	205	Opening of the Atlantic Ocean
Triassic	240	
Permian	290	Final stage in formation of Appalachian Mountains Extensive faulting
Pennsylvanian		
Mississippian	330 — 360	Second stage in formation of Appalachian Mountains, the continental collision with Africa
Devonian	410	
Silurian	435	Initial stage in formation of Appalachian Mountains Extensive metamorphism
Ordovician		
Cambrian	500 — 544	Deposition of carbonate rocks, found west of the park in Shenandoah Valley Deposition of siliciclastic rocks overlying volcanics
Precambrian	570 — 1000	Eruption of Catoctin volcanics Erosion of Grenville Mountains Formation of Grenville Mountains, probably similar in size to present-day Himalayas
	1200	Intrusion of granites and granodiorites, and deposition of sediments that later became the Pedlar and Old Rag formations

Run Formation, a discontinuous sequence of sandy conglomerates (see Stop 11 or 15).

Some time around 600 million years ago, the North American continent began to break apart from the land mass attached to its east side. Geologists believe that this land mass is now the Baltic Region of northern Europe. A large rift valley slowly opened, and at about 570 million years ago, basaltic magma derived from the Earth's mantle rose to the surface of the earth. The rift valley then flooded with lava that would later become the Catoctin metabasalts.

Volcanic rocks assigned to the Catoctin Formation extend from central Virginia to southern Pennsylvania. Similarly aged volcanic rocks occur in the Hudson Highlands of northern New Jersey and southern New York. There are also outcrops in northern Vermont, southern Quebec, and at the northern tip of Newfoundland. These lavas may have been extruded in a single province from a common source, like the vast outpourings of the Deccan Trap lavas, which covered one-fifth of India. It is also possible that each occurrence and locality represents a distinct and separate extrusive episode. Regardless of the extent of their distribution, the Catoctin magmas, so prominently displayed in Shenandoah National Park, are the largest vestige of this ancient volcanic event.

Volcanism of the Catoctin magmas may have lasted a few million years—perhaps ten or fifteen million—but probably no longer. Radiometric age determinations of more recent flood basalt provinces indicate a duration of only a few million years. When the Catoctin volcanic activity ceased, the rift basin continued to open and slowly subside (settle to a lower elevation). Rivers and streams flowed through it, leaving behind the deposits of sand and gravel that became the Weverton Formation. The land continued to subside to below sea level and salt water began to fill the basin.

The development of this ocean probably lasted several million years and deposited thick layers of sand along beaches, which the wind blew into large coastal dunes. Mud was deposited in lagoons and along tidal flats. The sand and mud were later compressed to form the sandstones, quartzites, and phyllites of the Hampton Formation, as well as the quartzites of the Erwin Formation.

The Erwin Formation is the youngest rock readily visible in Shenandoah National Park, but younger rocks exposed in the Valley and Ridge Province to the west tell us of subsequent geologic events. Thick sequences of Cambrian- to Ordovician-aged limestone represent the deposition of lime muds in an ocean that once covered the area. Limestones are deposited in warm, quiet, marine environments, so we know that Shenandoah National Park, during the late Cambrian to early Ordovician period was characterized by a warm, perhaps tropical climate. It may have been another Bermuda vacation spot!

The ocean basin began to close in middle Ordovician time as crustal plates ceased moving apart and began to move together. This happened in three pulses, each marked by one of the continental collisions that formed the Appalachian Mountains. The first collision occurred about 450 million years ago and produced the metamorphism that altered the rocks of Shenandoah National Park. Evidence suggests that the first event involved the collision of several island arc complexes with North America. These fragments of "foreign" land are now part of the North American continent, and some of them are found in the Piedmont Province. Today, a similar event would occur if North America and Asia began to move together, slowly closing the western part of the Pacific Ocean. This would cause the many islands and volcanic areas, including Japan, the Philippines, and Indonesia, to collide with and become attached to the Asian continent.

The second event was about 350 million years ago and involved the collision of Africa with North America. The third event, 300 million years ago, may have been a final pulse of the second event and involved the final docking of North America with Africa. It caused the rocks to be compressed and resulted in the creation of thrust faults, where large sections of continental crust were pushed up and thrust over rocks to the west. These faults can be found on the west side of the Blue Ridge Mountains (see figure 7, page 21, and photos 14-2 and 14-3) and on the west sides of many of the ridges in the Valley and Ridge Province.

These collisional events were part of the worldwide assembly of crustal fragments that led to the creation of *Pangea*, the supercontinent that included most of the planet's landmasses. This supercontinent had barely finished assembling when it began to break apart, about 200 million years ago. A rift formed between North America and Africa, quite close to where the two continents had been sutured together 100 million years earlier. This rift opened to form the Atlantic Ocean. As the rift formed, feeder dikes once again transported basaltic magma to the surface, and lava flowed into a large rift valley. These effects occurred primarily in the Piedmont area, east of the Blue Ridge, with only an occasional small intrusion in Shenandoah National Park.

Since then, uplift of the area during the Tertiary period and erosion have been the primary geologic agents that have altered the land. Gradual erosion of the hills and valleys by wind and rain has been the driving force, but from time to time massive and sudden landslides have moved large volumes of rocks and dirt from the flanks of the mountains to the valleys below. The most recent occurrence was in June 1995, following torrential rain storms, when slides occurred on the east side of the park in the drainage basins of the Staunton River, the Rapidan River, and the North Fork of the Moormans River.

New on the stage of geologic change is the role played by acid precipitation, which over the last few decades has altered the chemistry of the soils and water in the park (see Stop 21, page 73). This chemical change is also altering the natural balance of life that has evolved over the ten thousand years since the Ice Age, when northern glaciers (none in Virginia) retreated and the climate warmed. ■

Signal Knob Overlook

Mile 5.7

Park at the overlook, and after enjoying the view of Massanutten Mountain to the west, cross the road to examine the rocks. These represent two flows of the Catoctin metabasalt with a layer of sedimentary rock between them.

These massive, dark green metabasalts, now fading to a weathered greenish brown, represent two distinct lava flows separated by the sediment. The sedimentary layer, about eye level from the road, is 15 to 25 inches thick along its exposed section and consists of fine-to-coarse sandstone with pebbles of quartzite in the upper part of the layer.

One metabasaltic flow is at the base of the exposed rock in the roadcut. Originally, this was a lava flow that hardened to basalt. Because the original minerals in the basalt were altered by heat and pressure during the event

that formed the Appalachian Mountains (millions of years after the basalts erupted), they can no longer be called basalts. Instead we call them metabasalts, indicating that they have been metamorphosed. These metabasalts are grayish green with pistachio- to olive-green portions when the mineral epidote is abundant. Within the metabasalt are some rusty, weathered cavities, up to an inch long, that probably represent *vesicles* where gases escaped from the top of the flow after eruption.

The layer of sedimentary rock that overlies the lower metabasalt consists of

SOUTHBOUND

- Distance from North Entrance Station
  5.7 miles
- Next stop: Indian Run Overlook
  5 miles

NORTHBOUND

- Distance from Indian Run Overlook
  5 miles
- Distance to North Entrance Station
  5.7 miles
- Distance from Thornton Gap Entrance Station
  25.7 miles

1-1 Sedimentary layer (interbed) of conglomeratic sandstone between two distinct basaltic lava flows of the Catoctin Formation. The sandstone represents material deposited by rivers or streams during a lengthy pause between volcanic episodes. Arrows mark the upper and lower contacts of the sandstone.

1-2 Close-up of sedimentary interbed consisting of small pebbles of quartzite in a finer sandstone matrix. The knife, for scale, is at the contact with the overlying metabasalt.

SIGNAL KNOB

1-3 View to the northwest from Signal Knob Overlook looking toward Massanutten Mountain and, at its north end, Signal Knob. The ridge is composed of folded sedimentary rock younger than the rocks found in Shenandoah National Park. During the Civil War, General Stonewall Jackson posted Confederate scouts here to look for Yankee patrols and to communicate messages up and down the Shenandoah Valley.

shaley sandstone at its base; the remainder is made up of a conglomeratic sandstone (see photo 1-2). The rocks represent a sediment deposited by a river or stream along the surface of the underlying metabasalt. When volcanism occurred, a basaltic eruption was commonly followed by a period of quiescence, probably lasting a few hundred or a few thousand years. During that time, streams, rivers, or lakes deposited sediments over the now hardened basalt. The sediments were eventually compressed into sedimentary rock.

Overlying the sedimentary layer is a thick unit of metabasalt, representing another flow of lava that covered the sediments. This flow, several tens of feet thick, would have covered an area of several square miles. Massive cliffs of this metabasalt flow can be observed for the next 0.6 mile along the east side of Skyline Drive.

Looking west from the overlook, you can see the long, low ridgeline of Massanutten Mountain (more fully discussed at Hogback Overlook, Stop 5, page 24). This ridge is composed of sedimentary rocks that are more resistant to erosion than rocks in the adjacent Shenandoah Valley. Signal Knob, at the north end of Massanutten Mountain, was used as a lookout and signaling post for Confederate scouts during the Civil War. ■

SOUTHBOUND

- Distance from North Entrance Station
  10.7 miles
- Distance from Signal Knob Overlook
  5 miles
- Next stop: Range View Overlook
  6.5 miles

NORTHBOUND

- Distance from Range View Overlook
  6.5 miles
- Next stop: Signal Knob Overlook
  5 miles
- Distance from Thornton Gap Entrance Station
  20.7 miles

Indian Run Overlook
Mile 10.7

This stop is at a roadside outcrop of Catoctin metabasalt that contains vertical fractures called columnar joints. Park at the overlook and walk north along the road to observe these ancient volcanic features that are still preserved in the rocks.

Opposite the overlook, along the north end of the roadside outcrop, there is the finest example of *columnar jointing* in metabasalts seen along Skyline Drive. Other locations will be discussed at Stops 9B and 15, but these are along hiking trails.

Near Indian Run Overlook the columns are five- to six-sided with diameters of 6 to 8 inches, about the width of your hand. At other locations in Shenandoah National Park, columns are up to 30 inches in diameter. The columns rise more or less vertically from the ground (see photo 2-1), but some of them are gently curved (see

photo 2-2).

Columnar jointing is caused by the contraction of cooling magma. Upon cooling, the surface of a lava flow gradually contracts due to a decrease in volume as the liquid changes to a solid. As the magma contracts, cracks propagate outward from regularly spaced points; the angles between cracks are also fairly regular (see figure 6). This process is similar to the formation of mudcracks as mud dries out. In basalts, after the cracks are formed at the surface, they continue to grow below the surface as the magma cools and crystallizes at depth. The result is the

2-1 Columns of metabasalt of the Catoctin Formation. These columns were formed by cooling and contraction of the magma, as discussed in the text.

formation of five- to six-sided vertical columns that extend from the top to the bottom of the flow.

The columns of cracked basalt tend to form perpendicular to the cooling surface and are vertical if the lava flow is horizontal. However, if the underlying surface is not flat, then the two cooling surfaces—the cooling surface on top of the flow and the one between the flow and the ground—will not be parallel and the columns may curve. Columns may also curve due to slight forward movement of the flow during cooling. This outcrop displays some excellent examples of curved columns (see photo 2-2).

Several national parks in the West have well-known columnar-jointed basaltic flows, such as Sheepeater Cliffs in Yellowstone National Park and Devil's Tower National Monument in Wyoming. The ones in Shenandoah, at

2-2 Curved columns of metabasalt as viewed from the roadcut. The curvature may be due to slight movement of the lava after partial solidification, or it may be due to an irregular cooling surface, as discussed in the text.

570 million years old, are much older than their western counterparts and are still preserved despite alteration of the rocks by heat, pressure, and metamorphic fluids during formation of the Appalachian Mountains. ∎

Figure 6

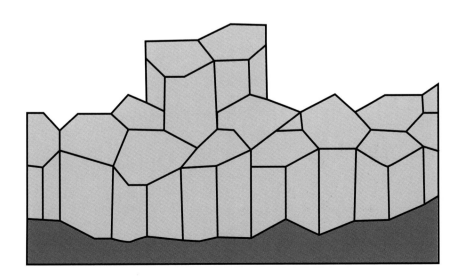

Five-to-six-sided columnar basalts formed by contraction during the cooling of magma.

Range View Overlook

Mile 17.2

This overlook offers an outstanding southward view of the crest of the Blue Ridge Mountains and is also one of the best and most accessible locations in the park to observe the contrast in physical appearance between the Piedmont, the Blue Ridge, and the Valley and Ridge provinces.

Before you is the most spectacular view of the Blue Ridge Mountains from Skyline Drive. From this viewpoint, several peaks, including Old Rag Mountain and Stony Man Mountain, can be identified up to 20 miles away on clear days. All of the peaks visible from the Range View Overlook are comprised of either the Catoctin Volcanics, the Pedlar Formation, or the Old Rag Granite.

This unique overlook along Skyline Drive offers views west and east of the Blue Ridge Mountains. To the west,

look for Massanutten Mountain, part of the the Valley and Ridge Physiographic Province. These ridges are several tens of miles long and are capped by resistant sedimentary rock, such as sandstone or conglomerate. The valleys between the ridges are composed of more erodable rock, such as shale or limestone. In some places, the ridges are formed by folds, places where the rocks were squeezed so that the rock layers buckled. In other places, the rocks have buckled and then broken along a gently dipping surface called a thrust

SOUTHBOUND

- Distance from North Entrance Station
  17.2 miles
- Distance from Indian Run Overlook
  6.5 miles
- Next stop: Little Devils Stairs Overlook
  2.9 miles

NORTHBOUND

- Distance from Little Devils Stairs Overlook
  2.9 miles
- Next Stop: Indian Run Overlook
  6.5 miles
- Distance from Thornton Gap Entrance Station
  14.2 miles

3-1 View to the south from Range View Overlook of the high ground constituting the spine of the Blue Ridge Mountains. The ravines between the peaks and ridges are the headwaters for the many mountain streams flowing from the mountains.

fault, causing the rocks on the ridge to be pushed to the west along the gently dipping fault surface, up and over the rocks in the adjoining valley (see figure 7).

To the east, the Piedmont Physiographic Province, which contains discontinuous, low-lying hills and ridges of old metamorphic rock, unfolds before you. Some of the rocks are part of the Catoctin Formation; others are metamorphosed sedimentary rocks that once filled rift valleys as the North American continent began to break apart during late Precambrian times, just before eruption of the Catoctin basalts. Geologists believe other rocks in the Piedmont formed elsewhere and became attached to the North American continent during a continental collision. Much work remains to unravel the complex geologic history of the Piedmont. ■

Schematic cross-section of the Valley and Ridge Physiographic Province showing the fault at the base of the Blue Ridge Mountains, the folded rock layers that form Massanutten Mountain, and the faults that underlie many of the ridges.

Figure 7

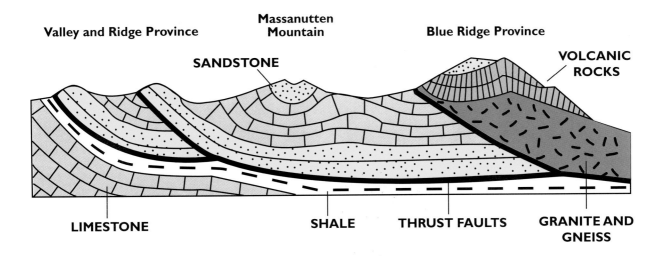

Valley and Ridge Province

Massanutten Mountain

Blue Ridge Province

SANDSTONE

VOLCANIC ROCKS

LIMESTONE

SHALE   **THRUST FAULTS**

GRANITE AND GNEISS

SOUTHBOUND

- Distance from
North Entrance Station
20.1 miles

- Distance from
Range View Overlook
2.9 miles

- Next stop:
Hogback Overlook
0.9 mile

NORTHBOUND

- Distance from
Hogback Overlook
0.9 miles

- Next Stop:
Range View Overlook
2.9 miles

- Distance from Thornton
Gap Entrance Station
11.3 miles

Little Devils Stairs Overlook
Mile 20.1

Several basaltic dikes are exposed along the rock face north of the overlook. Park at the overlook and walk about 300 feet north along the road to milepost 20.

This outcrop consists of coarsely crystalline, light gray gneiss of the Pedlar Formation (also seen at Stop 5, page 24). Just north of milepost 20, look for the five basaltic dikes that cut through the Pedlar Formation. The first dike is about 40 feet north of the milepost and the fifth is about 100 feet north. A sixth dike is about 270 feet north of the post (see photo 4-2). Fresh surfaces of the dikes are dark greenish gray, while the weathered surfaces are gray brown. The dikes are darker and much finer grained than the surrounding gneiss; they trend both west to northwest and east to northeast and are inclined steeply to the north. Thicknesses vary from several inches to nearly 10 feet.

Geologists interpret these dikes to be feeder dikes to Catoctin lava flows. Magma intruded along fractures in the Pedlar Formation, rising through the ancient granitic rock and extruding onto the surface of the 570-million-year-old landscape. At this location, the magma flows are now completely eroded; all that remains are the magmas which crystallized in the fractures that cut the Pedlar Formation.

4-1 Ledges of the Pedlar Formation, consisting here of medium-grained granodiorite. Six basaltic dikes that intrude the Pedlar can be seen at this location near Little Devils Stairs Overlook.

4-2 Basaltic dike intruding Pedlar Formation, located about 270 feet north of milepost 20. Such dikes clearly are younger than the rocks they cut. This dike could be called a "feeder dike," since we believe that lava once flowed through it to the Earth's surface to form a basaltic lava flow of the Catoctin Formation. Notebook, for scale, is 5" x 7.5".

When Tom Gathright of the Virginia Division of Mineral Resources mapped the geology of the park during the 1970s, he located more than one hundred of these dikes, including those at this stop. ■

SOUTHBOUND

- Distance from
  North Entrance Station
  21 miles

- Distance from
  Little Devils Stairs
  Overlook
  0.9 miles

- Next stop:
  Thornton River Trail
  Parking Area
  4.4 mile

NORTHBOUND

- Distance from
  Thornton River Trail
  Parking Area
  4.4 miles

- Next Stop:
  Little Devils Stairs
  Overlook
  0.9 miles

- Distance from Thornton
  Gap Entrance Station
  10.4 miles

Hogback Overlook

Mile 21

GEOLOGY 5 STOP

This is a long overlook with ledges of the Pedlar Formation on the east side of the road and views of the Valley and Ridge Province on the west side.

Stand at the overlook and take a good look at the Valley and Ridge Province so well displayed to the west. The long, linear ridges are composed of sedimentary rocks that are mostly 350 to 450 million years old, approximately one to two hundred million years younger than the metabasalt of the Catoctin Formation (Stops 1 and 2). The Valley and Ridge Province is composed of sedimentary rocks that were deposited in the ocean basin that covered the area after the volcanic rocks flowed over the land. The tops of the ridges are made up of harder sedimen

tary rock, usually sandstone or conglomerate, while the valleys are composed of softer rock that erodes more easily, such as shale or limestone. On a clear day, at least five ridges are visible from Hogback Overlook, one behind the other.

Now look behind you at the rock outcrops on the other side of the road. These old metamorphic rocks are part of the Pedlar Formation, which underlies the Catoctin volcanic flows seen at Stops 1 and 2. They are tan to light gray with interlocking crystals, up to an inch in diameter, of light gray to clear quartz,

5-1 Massanutten Mountain and ranges of the Valley and Ridge Province as seen to the west from Hogback Overlook. In the foreground is the South Fork of the Shenandoah River as it meanders north through the Shenandoah Valley.

◄ ▮▮▮

5-2 Roadside outcrop of Pedlar Formation at Hogback Overlook. These are primarily granodiorite, composed of quartz and feldspar.

◄ ▮▮▮

5-3 Granitic gneiss of the Pedlar Formation. The cream-colored mineral is feldspar (both plagioclase and orthoclase feldspar); the gray is quartz. Note the banding formed by alignment of the quartz grains. It is this banding that allows geologists to classify this rock as a gneiss. Nickel on the upper left is for scale.

cream-colored orthoclase feldspar and plagioclase feldspar. The few darker-colored minerals are pyroxene and hornblende. The rocks are a type called granodiorites, which are similar to granites in that they consist primarily of quartz and two types of feldspar. However in granodiorites, calcium-rich feld-spar (plagioclase) predominates over potassium-rich feldspar (orthoclase), while in granites the two feldspars are in approximately equal concentrations. These rocks are between 1 and 1.2 billion years old.

The Pedlar Formation is well displayed in the outcrops along the

5-4 Stone wall at Hogback Overlook, composed of gneiss and granodiorite of the Pedlar Formation. These rocks were probably blasted from the outcrop along the road during the construction of Skyline Drive in the 1930s. The rocks were trimmed with chisels to construct the wall.

roadcut opposite the overlook. It is also visible in the rock wall in front of you at the overlook, which was constructed from rocks of the Pedlar Formation. Look for several varieties of the Pedlar Formation in the wall—gray rocks containing abundant quartz, light-colored rocks containing mostly feldspar, darker-colored rocks containing considerable amounts of pyroxene and hornblende, along with a few rocks that are dark red. The latter rocks reflect the presence of hematite, an oxidized form of iron, formed by the weathering of iron-bearing minerals.

For a view of the Pedlar Formation uncluttered by auto traffic, go a few hundred feet south of Hogback Overlook to the junction with the Appalachian Trail. There is a parking area on the west side of the road, but you can easily walk from the overlook. From the east side of Skyline Drive, hike between 700 and 800 feet along the Appalachian Trail to a rise overlooking the Hogback Overlook parking area. The view is just as spectacular from here, but it is quieter and you do not have to dodge the cars. ■

Thornton River Trail Parking Area

Mile 25.4

Metasediments overlying Catoctin metabasalts can be seen along the Thornton River Trail as it heads west from Skyline Drive. From the parking lot, walk south about 150 feet along Skyline Drive to the Thornton River Trail heading west.

At the beginning of the trail, there are 19 rock steps, consisting primarily of Catoctin metabasalt and rocks of the Pedlar Formation, which have been transported from elsewhere. Above these stairs and continuing uphill, look for a good sampling of the rock types in the Weverton and Hampton formations. They include black shales; tan, brown, gray, red, and green siltstones and sandstones; red *argillaceous* conglomerates; and coarse quartz-pebble conglomerates. A few outcrops of black shale and tan sandstone appear on the uphill side of the trail. All of the rock varieties mentioned above can be seen when hiking the first 400 feet along the trail.

These rocks were sediments deposited on top of the volcanic rocks of the Catoctin Formation. After volcanic activity ceased in the Shenandoah area, the continent continued to pull apart, the land slowly subsided to a lower elevation, and an ocean basin began to form between the two separating land masses. This was not the Atlantic Ocean, which did not begin to open

SOUTHBOUND

- Distance from North Entrance Station
  25.4 miles
- Distance from Hogback Overlook
  4.4 miles
- Distance to Thornton Gap Entrance
  6 miles
- Next point of interest (don't stop): Marys Rock Tunnel
  6.8 miles

NORTHBOUND

- Distance from Marys Rock Tunnel
  6.8 miles
- Next Stop: Hogback Overlook
  4.4 miles
- Distance from Thornton Gap Entrance Station
  6 miles

6-1 Samples of the sedimentary rocks that overlie the Catoctin metabasalts. These sedimentary rocks represent deposition in Cambrian-aged lakes, rivers, and along marine shorelines. Rock varieties include: quartz-pebble conglomerate (top), tan sandstone (right), black phyllite (bottom), gray sandy siltstone (left), and red sandstone (middle).

until about 200 million years ago, but was an ocean basin long since gone called the Iapetus, or proto-Atlantic, Ocean.

A wide variety of sediments were deposited in this new, subsiding basin. Silt and mud accumulated in shallow lagoons and lakes; conglomerates and sandstones were deposited by rivers; more sandstones were deposited as beach sands as the sea slowly enveloped the area. These sediments were later compressed to form the sedimentary rocks of the Weverton and Hampton formations, which overlie the Catoctin metabasalts in this part of Shenandoah National Park. These rocks tend to weather fairly rapidly and do not readily form outcrops here. Most of the exposed rock here is what geologists call *float,* loose rock on the surface of the ground that has broken off or been weathered from outcrops that now lie just below the surface of the ground. This is the best place in the northern part of the park to see these rock types.

An alternative stop is at Jeremys Run Overlook, mile 26.4. Small outcrops of tan sandstone from the Weverton Formation are exposed on the bank opposite the overlook, but no other rock types are exposed. ■

Marys Rock Tunnel

Mile 32.2

Do not stop here, as there is no place to park, but drive by slowly for a look at the basaltic dike next to the north end of the tunnel.

At the north end of the tunnel, on the west side of the road (right side for southbound travelers), look for the 12-foot-wide basaltic dike that cuts across the gneiss of the Pedlar Formation. This was one of the feeder dikes to the Catoctin lava flows. Not all dikes display vertical intrusion, as demonstrated by this one, which dips about 68 degrees to the northwest, with a northeast trend. ■

SOUTHBOUND

- Distance from Thornton Gap Entrance Station
  0.8 miles

- Distance from Thornton River Trail Parking Area
  6.8 miles

- Next stop: Hazel Mountain Overlook
  0.9 mile

NORTHBOUND

- Distance from Hazel Mountain Overlook
  0.9 miles

- Distance to Thornton Gap Entrance Station
  0.8 miles

- Next Stop: Thornton River Trail Parking Area
  6.8 miles

- Distance from Swift Run Gap Entrance Station
  33.3 miles

Inclined 12-foot-wide dike of Catoctin metabasalt that cuts the Pedlar gneiss on the north side of Marys Rock Tunnel.

SOUTHBOUND

- Distance from Thornton Gap Entrance Station
  1.7 miles

- Distance from Marys Rock Tunnel
  0.9 mile

- Next Stop: Pinnacles Overlook
  2 miles

NORTHBOUND

- Distance from Pinnacles Overlook
  2 miles

- Next point of interest, but don't stop: Marys Rock Tunnel
  0.9 mile

- Distance from Swift Run Gap Entrance Station
  32.4 miles

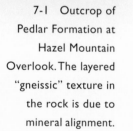

7-1 Outcrop of Pedlar Formation at Hazel Mountain Overlook. The layered "gneissic" texture in the rock is due to mineral alignment. Garnets are peppered throughout the outcrop and are best observed on the rock slabs facing to the right (east).

Hazel Mountain Overlook
Mile 33.1

The outcrop at the overlook is gneiss of the Pedlar Formation, containing small crystals of dark red garnet.

The Pedlar Formation is well exposed at this overlook. The rock is light-colored, consisting mainly of feldspar and quartz, but also peppered with minute grains of dark red garnet. The garnets are 1 to 3 mm in diameter and are best seen on the side of the outcrops that face away from the parking area. Photo 7-2 is a photograph of one of these garnets taken through a microscope. Also, there is a pronounced layering in the rock, called *gneissic texture*, formed by the alignment of minerals along planar surfaces. The minerals, primarily quartz, appear light gray in the outcrop. The grains aligned themselves during metamorphism when the rocks were heated and squeezed, causing the existing minerals to recrystallize. Garnet was also formed during this metamorphic event.

Look east from the overlook for a good view of the Piedmont Province and the low-lying hills of ancient metamorphic rock that rise from it (see photo 7-3). ■

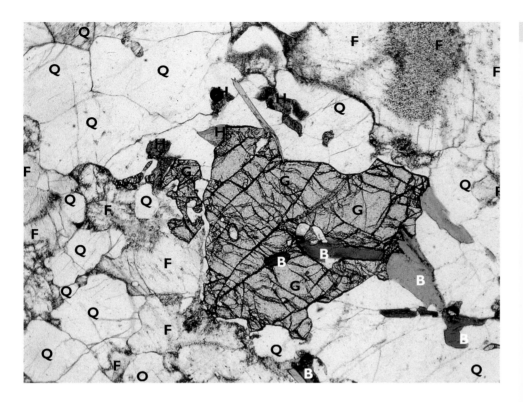

7-2 Microscopic view of thin section of Pedlar Formation showing garnet with feldspar, quartz, biotite, and a little hornblende. Field of view is 3 mm.

G = garnet
B = biotite
Q = quartz
F = feldspar
H = hornblende

7-3 View east from Hazel Mountain Overlook. The low, irregular hills are composed of metamorphic rock that comprise the Piedmont area of Virginia. An early morning ground fog highlights the hills.

- Distance from Thornton Gap Entrance Station
  3.7 miles

- Distance from Hazel Mountain Overlook
  1.9 miles

- Next stop: Stony Man Overlook
  3.5 miles

- Distance from Stony Man Overlook
  3.5 miles

- Next Stop: Hazel Mountain Overlook
  1.9 miles

- Distance from Swift Run Entrance Station
  30.4 miles

Pinnacles Overlook

Mile 35.1

From this overlook you can view the profile of Old Rag Mountain, which is located in an area of Shenandoah National Park that is not readily accessible from Skyline Drive.

Old Rag Mountain is composed of granite—called, appropriately enough, the Old Rag Granite—which differs slightly in composition from the granodiorite of the Pedlar Formation. Old Rag Granite contains more orthoclase feldspar than the granodiorite of the Pedlar Formation and is also coarser grained. The quartz grains have a blue-gray color, and veins of blue-gray quartz cut the granite. Of prime interest on Old Rag are several basaltic dikes, feed-ers to Catoctin flows, which cut the granite (see photo 8-2). These dikes erode more easily than the granite, so their removal has left crevasses in the granite. In two places the hiking trail passes through these crevasses.

This is a wonderful mountain to climb, but the hike takes most of a day. To get to it, you must leave the park and drive to the trailhead or hike several miles to the trailhead via the Corbin Hollow Trail, Nicholson Hollow Trail,

8-1 Old Rag Mountain as seen from Pinnacles Overlook. The trail to the top leads from the left (north) along the rocky ridgeline to the summit at 3,268 feet above sea level.

8-2 Three-foot-wide crevasse, eroded within a tabular basaltic dike, through which the trail passes leading to the top of Old Rag Mountain. The floor consists of dark, fine-grained basalt, whereas the steep walls consist of coarse-grained crystalline Old Rag Granite.

or Old Rag Fire Road. More detailed directions and a guide to the geology are offered in Appendix 1 for those who have the time to climb.

There are also good views of Old Rag from Thorofare Mountain Overlook, mile 40.6; at Old Rag View Overlook, mile 46.5; and from the top of Hawksbill Mountain (Stop 11). ▪

SOUTHBOUND

- Distance from Thornton Gap Entrance Station
  7.2 miles
- Distance from Pinnacles Overlook
  3.5 miles
- Next stop: Little Stony Man Parkiing Area
  0.5 mile

NORTHBOUND

- Distance from Little Stony Man Parking Area
  0.5 miles
- Next Stop: Pinnacles Overlook
  3.5 miles
- Distance from Swift Run Entrance Station
  26.9 miles

Stony Man Overlook
Mile 38.6

Erosion of the metabasalts on the mountain southwest of this overlook has produced an outline of the face of a man, hence the name: Stony Man Mountain.

From the parking area, try to make out the features of the skyward-looking face on the west side of the mountain. The smoothly curving forehead at the top of the mountain breaks to mark the indentation for the eyes. A large blunt nose is followed by a long flowing beard. The features that form the face result from weathering along the contacts between individual metabasaltic flows of the Catoctin Formation (see photo 9A-1 and figure 8). There are five metabasalt flows on the mountain, overlying rocks of the Pedlar Formation. The bottom three flows form the beard. The junction between the third and fourth flows—the base of the nose—is marked by a volcanic *breccia* that should not be missed; go to Stop 9B and hike the trail to Little Stony Man, where this breccia can be seen. The fourth flow forms the nose and the fifth flow forms the forehead. These flows are no longer horizontal but dip gently to the east.

The features of the face are highlighted by weathering along the contacts between the different metabasaltic

9A-1 Stony Man Mountain as seen from Stony Man Overlook. Note the rounded forehead, the break for the eyes, the long nose, then mouth and flowing beard. The breaks for the eyes and below the nose are formed by erosion along the contact zones between metabasaltic flows.

34

flows. This is because the composition of these contact zones is often different from that of the flows themselves, sometimes containing volcanic breccia (see Stop 9B, page 36) or sedimentary interbeds (see Stop 1, page 15). The result is that the contact zone forms a natural break during weathering, thus creating the step-like breaks frequently found between flows throughout the park, including here at Stony Man. ■

Figure 8

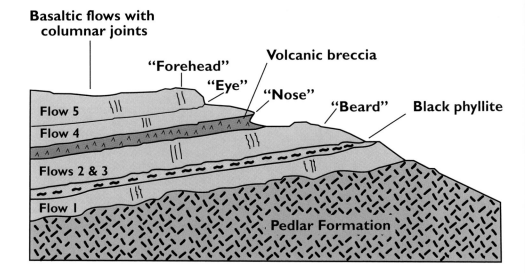

Basaltic flows with columnar joints

"Forehead"

"Eye"

"Nose"

Volcanic breccia

"Beard"

Black phyllite

Flow 5

Flow 4

Flows 2 & 3

Flow 1

Pedlar Formation

Sketch of Stony Man showing the metabasaltic flows and the breaks between them that form the features of the face.

3 5

SOUTHBOUND

- Distance from Thornton Gap Entrance Station
  7.7 miles

- Distance from Stony Man Overlook
  0.5 miles

- Next stop: Crescent Rock Overlook
  5.2 mile

NORTHBOUND

- Distance from Crescent Rock Overlook
  5.2 miles

- Next Stop: Stony Man Overlook
  0.5 miles

- Distance from Swift Run Entrance Station
  26.4 miles

Little Stony Man Parking Area
Mile 39.1

Hike 1 to 1.5 miles to see the volcanic breccia between two metabasaltic flows on Little Stony Man Mountain.

The trail to Little Stony Man begins at this parking lot. Head up the trail for 0.4 mile to a trail junction and stone marker. The white-blazed Appalachian Trail turns left (south) to Little Stony Man, but continue straight ahead on the blue-blazed Passamaquoddy Trail for about 600 feet to the first good viewpoint. This viewpoint is at the contact between the third and fourth layers of the Catoctin volcanic rocks. These two layers are separated by a zone of volcanic breccia, a large block of which is on your left (see photo 9B-1).

Contained in the breccia are many fragments of what appear to be dark red shales or siltstones within a metabasalt matrix (see photo 9B-2).

There are two theories as to how these fragments were formed. One is that muds and silts were deposited in a shallow lagoon or lake on top of the third lava flow. The sediments provided a base upon which the fourth lava flow was extruded, and pieces were ripped up as it advanced over the area. The second theory is that the top of the third flow hardened while its center remained

9B-1 Large boulder of volcanic breccia at break between the third and fourth flows of the 570-million-year-old Catoctin volcanic sequence. This breccia contains abundant fragments of sedimentary rock within a matrix of volcanic lava. 13-year-old Dan Badger for scale.

◀ ▪▪▪

9B-2 Close-up of a reddish siltstone fragment within the volcanic breccia. The fragment was probably from an unconsolidated sedimentary layer deposited on top of the third flow that was ripped up and included in the breccia as the fourth flow moved over the area.

◀ ▪▪▪

9B-3 A Brunton compass, companion of every field geologist, showing magnetization of the third volcanic flow. The black arm points to the right toward true north, while the white point of the compass needle is deflected by magnetite in the rock.

molten. As the molten part continued to move, the top fractured into irregular, broken blocks of basalt, producing the very irregular surface commonly seen on volcanic flows today. Then, as wind-blown or water-transported sediments entered the area, they filled in around the irregular basalt blocks to deposit pockets of sediments, similar to the observed red siltstone fragments.

These red siltstone fragments are common within breccias of the Catoctin volcanics, and the occasional presence of individual siltstone beds between flows makes both theories quite plausible. Because the siltstone fragments are fairly small and quite abundant, and because the orientation of the layering in them varies from fragment to fragment, I favor the former hypothesis, that these were fragments of unconsolidated sediment that were ripped up by an advancing lava flow.

Also of interest here is the magnetism of the rocks at the top of the third flow. The iron oxide mineral *magnetite* is so abundant here that it deflects the needle of a compass (see photo 9B-3).

Just beyond the overlook, the trail runs along the base of the vertical wall of the fourth flow, which is approximately 95 feet thick. Large columns of metabasalt are visible at the base of this flow, rising to the top of the cliffs above (see photo 9B-4).

From the overlook, return 600 feet to the junction with the Appalachian Trail and follow it to the top of Little Stony Man, a distance of about 0.25 mile. You are now standing at the top of the fourth flow, and can peer over the edge of the near-vertical, 95-foot wall. From this location you can also trace the lava flow upward, at an 8-degree slope, to the break between flows that forms the eyes in the profile of Stony Man's face.

The two viewpoints from Little Stony Man offer excellent views of the Blue Ridge Mountains to the north (see back cover) and of Massanutten Mountain to the west. Directly west, there is a break in the ridge called New Market Gap, through which General Stonewall Jackson marched his troops on several occasions during the early part of the Civil War. See the discussion of Jackson's campaign in the Shenandoah Valley (page 46). ■

9B-4 View of columnar metabasalts composing the fourth flow looking up from the top of the third flow. The vertical columns were produced by shrinkage during the cooling of liquid to solid basalt. Columns are about 30 inches in diameter and 95 feet tall.

Crescent Rock Overlook
Mile 44.3

This overlook is on top of volcanic rocks of the Catoctin Formation and provides fine views of Hawksbill Mountain to the southwest.

SOUTHBOUND

- Distance from Thornton Gap Entrance Station
  12.9 miles
- Distance from Little Stony Man Parking Area
  5.2 miles
- Next stop: Hawksbill Gap Parking Area
  1.4 mile

NORTHBOUND

- Distance from Hawksbill Gap Parking Area
  1.4 miles
- Next Stop: Little Stony Man Parking Area
  5.2 miles
- Distance from Swift Run Entrance Station
  21.2 miles

Crescent Rock is located on the third lava flow from the base of the Catoctin volcanic sequence. Columnar jointing is visible here, but it is poorly preserved. Unlike many of the volcanic flows, no volcanic breccia is found at the top of this flow. Instead, this stop provides you with an opportunity to see the inside or central portion of a flow. The erosion surface cuts diagonally across the flow at a gentle enough angle, so you can climb down the cliffs (there are even a few stairs cemented into place) and observe a cross section of the flow from the top to near the bottom. The flow is thick and massive throughout, with remarkable consistency. It is not recommended to climb on the cliffs on windy days, so if the wind is howling, park at the overlook for a few minutes and watch, as ravens frequently can be seen playing in the updrafts along the cliffs.

Crescent Rock is an excellent place from which to view Hawksbill Mountain, the highest mountain in the park at 4,049 feet above sea level. The base of Hawksbill Mountain consists of the

10-1 Metabasaltic cliffs at Crescent Rock as viewed from the trail to Hawksbill Mountain. The eroded surface of the cliff allows you to see the uniformity of the rock from the upper surface of the flow to its center.

10-2 Hawksbill Mountain and cliffs of the metabasaltic flows as seen from Crescent Rock. Exposures of several of the five flows can be seen, and a sixth thinly caps the mountain just to the left (east) of its top. Refer to Figure 9 to see the configuration of the volcanic flows.

Figure 9

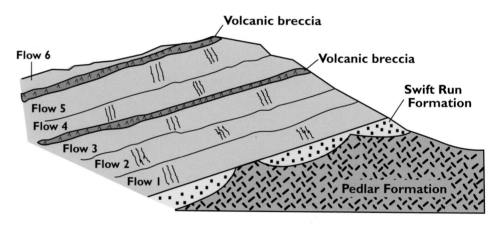

Volcanic breccia

Volcanic breccia

Flow 6

Swift Run Formation

Flow 5

Flow 4

Flow 3

Flow 2

Flow 1

Pedlar Formation

Sketch of Hawksbill Mountain showing the six flow units of metabasalts, the breccias between some of the flows, the discontinuous Swift Run Formation beneath the flows, and the Precambrian gneiss that underlies the mountain.

Pedlar Formation, while the younger rock at the top is the fifth flow of the Catoctin volcanic sequence (see photo 10-2 and figure 9). Intervening and forming several cliffs are four other Catoctin volcanic flow units, plus a thin layer of sedimentary material between the Pedlar and the Catoctin. A sixth flow, the youngest, occurs just east of the mountaintop (see figure 9). The flows are best seen when the leaves are off the trees, but even in summer, cliffs marking the flows are visible. Blocks and boulders of metabasalt have eroded from the cliffs and fallen into piles below called *talus slopes* (see figure 10, page 42). View the lava flows from here, and then, if time permits, drive to Hawksbill Gap Parking Area and hike the 2.9-mile loop trail that goes around, up, over, and back down the mountain. ■

Hawksbill Gap Parking Area
Mile 45.7

Park here and hike the 2.9-mile loop trail that goes over Hawksbill Mountain. At 4,049 feet above sea level, Hawksbill is the tallest mountain in Shenandoah National Park. The mountain is easily accessible from the road by several trails, since Skyline Drive is above 3,000 feet in this part of the park.

From the trailhead at Hawksbill Gap Parking Area, take the short connecting trail to the Appalachian Trail and hike south approximately 0.4 mile to the first talus slope. The talus is composed of fragments of basalt derived from the cliffs above. This material is constantly being weathered and broken from the cliffs by the expansion of water to ice in winter and thawing in spring (figure 10). Rattlesnakes like to den in talus slopes such as these.

The trail, where it crosses the 250-foot-wide talus slope, offers good views of Crescent Rock to the north. Note the tops of the ledges and remember from discussions at previous stops that these usually represent the tops of flows. From this vantage point, the two volcanic flows below Crescent Rock can be delineated, as well as three flows on the hill above and across the road from Crescent Rock.

About 100 feet beyond the first talus slope, look for an outcrop of the Swift Run Formation (see photo 11-1),

SOUTHBOUND

- Distance from Thornton Gap Entrance Station
  14.3 miles
- Distance from Crescent Rock Overlook
  1.4 miles
- Next Stop: Franklin Cliffs Overlook
  3.3 miles

NORTHBOUND

- Distance from Franklin Cliffs Overlook
  3.3 miles
- Next Stop: Crescent Rock Overlook
  1.4 miles
- Distance from Swift Run Entrance Station
  19.8 miles

11-1 Lichen-covered outcrop of Swift Run Conglomerate along the Appalachian Trail on northwest side of Hawksbill Mountain. The conglomerate consists of pebbles of quartzite, up to 1 cm in diameter, in a matrix of coarse sandstone. This material was probably deposited on the pre-volcanic landscape by a stream.

11-2 Talus slopes on northwest side of Hawksbill Mountain. This material is broken off from the cliff faces during thawing and freezing cycles, and is slowly transported downslope by frost action.

the metasedimentary rock unit underlying the Catoctin volcanics. The Swift Run is composed of coarse conglomeratic sandstone, which contains quartzite pebbles up to half an inch in diameter, along with some sandy phyllite.

Outcrops and broken-off blocks of conglomerate continue for the next 180 feet along the east (left) side of the trail, with an impressively huge boulder at a distance of 180 feet from the first Swift Run outcrop. Continuing south along

Typical weathering of the metabasaltic cliffs. Frost action causes rocks to erode from the cliff faces, forming talus slopes below.

Figure 10

11-3 Catoctin metabasalt on top of Hawksbill Mountain, the upper surface of the fifth metabasaltic flow. Peregrine falcons were released from here during the early 1990s.

the Appalachian Trail, you cross the Swift Run conglomerate and walk back onto the Catoctin metabasalt, as evidenced by the proximity of metabasaltic cliffs to the left of the trail. Eventually, the trail crosses the lowermost metabasalt flow.

One mile from the parking lot at Hawksbill Gap, there is a junction with a blue-blazed trail to the left. This trail heads south, then east, then northeast 0.9 mile up the gentle southwest flank of Hawksbill Mountain to the peak, which is on the top of the fifth metabasalt flow. As with the third flow at Stony Man, parts of this flow contain so much magnetite that the needle of a compass is deflected. Columnar jointing of the fifth flow is marginally preserved and can still be seen in places.

In the woods and along the trails east of the peak, notice the outcrops and

float, or loose blocks of volcanic breccia, that cap the fifth flow. The volcanic breccia's resistance to weathering is largely responsible for the preservation of the peak.

Peregrine falcon chicks were raised and released during the early 1990s from the top of Hawksbill Mountain in an effort to reintroduce this species into the park. In subsequent years, some have returned and are sometimes seen from the top of Hawksbill Mountain.

From the top of Hawksbill Mountain, a blue-blazed trail leads 0.9 mile back to Hawksbill Gap Parking Area. ■

- Distance from Thornton Gap Entrance Station
  17.6 miles

- Distance from Hawksbill Gap Parking Area
  3.3 miles

- Next stop: Dark Hollow Falls Parking Area
  1.6 mile

NORTHBOUND

- Distance from Dark Hollow Falls Parking Area
  1.6 miles

- Next Stop: Hawksbill Gap Parking Area
  3.3 miles

- Distance from Swift Run Entrance Station
  16.5 miles

Franklin Cliffs Overlook

Mile 49

Franklin Cliffs Overlook presents an outstanding example of volcanic breccia. Some of the rock fragments in the breccia look similar to the conglomerates of the Swift Run Formation, which can be seen along the trail to Hawksbill Mountain at Stop 11.

Franklin Cliffs are composed of lava from the fifth flow in the Catoctin metabasalt sequence. As is common in the park, metabasaltic breccia, consisting of large, blocky fragments of basalt, granite, and conglomerate, caps the flow. The breccia has a basaltic matrix that has been altered mostly to the pistachio- to olive-green–colored mineral epidote. The examples of metabasaltic breccia at this stop are outstanding and easily accessible.

Outcrops of breccia can be found in the woods between the overlook and Skyline Drive, but the best example is the large boulder at the north end of the overlook (see photo 12-1). Some of the rock fragments within the breccia are made up of conglomeratic material very similar to the rock types found in the Swift Run Conglomerate (see photo 12-2). If these are indeed from the Swift Run Formation, they provide evidence that at the time of volcanic activity, the landscape had not been eroded to a flat plain but instead was hilly. While ba-

12-1 Boulder of metabasaltic breccia. See close-up on next page. 8-year-old Dylan Badger is shown for scale.

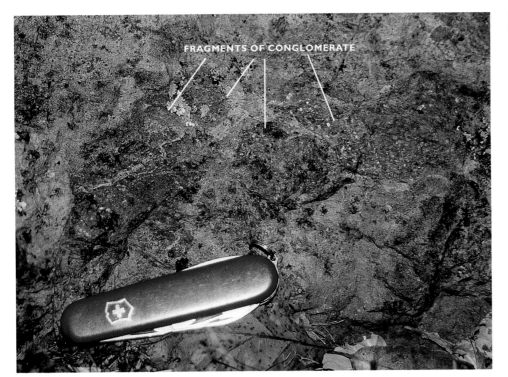

FRAGMENTS OF CONGLOMERATE

◀ ▮ ▮ ▮

12-2 Metabasaltic breccia containing angular fragments of conglomerate that appear to be from the Swift Run Formation. This material is similar to conglomerate of the Swift Run Formation seen at Stop 11.

◀ ▮ ▮ ▮

12-3 Top of Franklin Cliffs looking into Fishers Gap through which General Stonewall Jackson led his troops in November 1862.

salt flowed through the valleys and lower elevations, fragments of eroded older rocks on the hillsides were incorporated into the flows, perhaps as the lava pushed out of fissures on the side of the mountains and down into the valleys.

The hill to the southwest is the location of Big Meadows Lodge and Campground. The lodge and surrounding cabins are not visible from here, though their lights can be seen at night. The lodge and the cabins are located on top of the tenth flow in the Catoctin volcanic sequence.

From this vantage point, you can trace the route of General Stonewall Jackson, who led 25,000 troops in November, 1862, from Antietam, Maryland, to Fredericksburg, Virginia, to help General Robert E. Lee repel a federal attack on Richmond, Virginia, the Confederate capital. The army crossed Massanutten Mountain through New Market Gap, visible from this stop, and crossed the Shenandoah Valley near the town of Stanley, also visible from here. The troops climbed the Blue Ridge Mountains to cross at Fishers Gap, a few hundred yards to the south. ■

12-4 Mountain laurel clinging to the sides of metabasalts of the Catoctin Formation at Franklin Cliffs.

Dark Hollow Falls Parking Area

Mile 50.6

A 2-mile round-trip hike leads from the parking lot to the waterfalls in Dark Hollow. The waterfalls cascade over metabasalts of the Catoctin Formation. In the streambed above the falls, look for an interesting conglomerate and zones of amygdules within the metabasalts; below the falls, notice the very unusual sandstone dikes in volcanic breccia.

This is the most accessible of Shenandoah's many waterfalls. Where waterfalls such as this occur, the volcanic flows of the Catoctin Formation are stacked in a layered sequence with surfaces that dip gently to the east. In the upper reaches of a stream valley, these surfaces provide inclined spillways for running water until some semblance of a stream channel is established. The stream continues to flow along the surface that marks the boundary between metabasalt layers until at some point the flowing water erodes through the upper surface and through the volcanic layer to the top of the underlying layer. If this erosion surface across the metabasalt layer forms an escarpment, a waterfall is created (see figure 11). Waterfalls along Whiteoak Canyon,

SOUTHBOUND

- Distance from Thornton Gap Entrance Station
  19.2 miles
- Distance from Franklin Cliffs Overlook
  1.6 mile
- Next Stop: Blackrock at Big Meadows Area
  0.6 mile on Skyline Drive, about 1 mile on access road

NORTHBOUND

- Distance from Big Meadows on Skyline Drive
  0.6 miles
- Next Stop: Franklin Cliffs Overlook
  1.6 miles
- Distance from Swift Run Gap Entrance Station
  14.9 miles

13-1 Upper falls at Dark Hollow Falls. The stream has flowed along the top of a metabasaltic flow for a considerable distance before eroding and cutting across the face of the flow to form the falls.

Figure 11

Formation of waterfalls by stream erosion. The stream has eroded from the top of one flow unit to the next unit down in the sequence.

13-2 Outcrop of Catoctin Formation with rounded clasts of Pedlar Formation and zones of amygdules, along Hogcamp Branch above Dark Hollow Falls.

Rose River, Doyles River, Jones Run, and here in Dark Hollow were all formed in this way.

In addition to the waterfalls, there are three interesting geologic features along the trail above and below Dark Hollow. The first is a conglomerate at the top of a metabasalt layer, the second is the presence of well-displayed *amygdules*, which occur sporadically within the metabasalts, and the third consists of small sandstone dikes found in some of the breccia zones.

The conglomerate is about 10 to 20 feet thick and occurs along a considerable distance of the central portion of

13-3 Conglomerate in the Catoctin Formation containing rounded fragments of gneiss from the Pedlar Formation. These rounded fragments, more resistant to erosion than the metabasalt within which they were emplaced, were probably stream gravels picked up by a lava flow.

13-4 Single rounded fragment of gneiss in conglomerate of the Catoctin Formation. Note faint layering in the fragment (highlighted by solid lines) which has a different orientation than the layering in the metabasalt (also highlighted by dashed lines), indicating it was pre-existing before the volcanic event.

13-5 Amygdules in metabasalt. These were air bubbles in the lava that later filled with quartz, feldspar, epidote and/or hematite. They are more resistant to erosion than the metabasalt, so they have a knobby appearance.

13-6 Sandstone dike in metabasaltic breccia near base of the upper falls. These may indicate sediments that filled in around blocks of hardened basalt near the top of a basaltic flow after the surface had crystallized and fractured. Nickel is shown for scale.

the trail (see photos 13-2 and 13-3). It consists of cobbles of quartzite and gneiss dispersed in a fine-grained metabasaltic matrix that may be either part of a volcanic flow or volcanic material reworked by streams (*volcaniclastic*). The gneiss cobbles are well rounded, as though they were deposited in a stream, and appear to be rounded cobbles of gneiss of the Pedlar Formation. In some cobbles, the pre-Catoctin fabric of the Pedlar is preserved (see photo 13-4). Thus, this conglomerate could be a stream deposit containing fragments of the Pedlar Formation eroded from some nearby highlands that became mixed with finer volcaniclastic material, or it could be the result of a volcanic flow that incorporated the stream-derived cobbles as it flowed over them. I tend to favor the latter theory, since amygdules occur between the cobbles (see discussion below of amygdules) which would only be expected if this material was a volcanic flow rather than reworked volcanic material. This is the only place in the park where I have found this conglomerate in metabasalt.

Amygdaloidal basalts are common in volcanic regions. Amygdules are areas within magmas, commonly near the tops of flows (sometimes at the very bottom) where gases, primarily water vapor and carbon dioxide, have escaped from the molten magma. As the gases bubble out of this viscous, molten goo, they create small air pockets, called vesicles, which become holes when the magma hardens. Later, the holes become filled with minerals precipitated by hot fluids moving through the solidified rocks. In this case, the vesicles are filled with quartz, epidote, albite, and hematite. These amygdules are more resistant to weathering than the basalt, so on some surfaces they stand out as white or cream-colored bumps, 2 to 5 mm in diameter (see photo 13-5). Zones of amygdules in the Dark Hollow Falls area are visible in the conglomerate described above, in the streambed just above the upper falls, and just below the base of the upper falls where the trail intersects the stream.

Small, contorted dikes of dark red siltstone occur in the volcanic breccia below the base of the upper falls in Dark Hollow (see photo 13-6). These dikes are unusual in that we usually think of dikes as magma-filled fractures in rock. In this case, however, sedimentary material has acted as though it had been a liquid, filling voids between angular blocks of basalt.

The material in the dikes is similar to the red, cherty siltstone found in the volcanic breccia between the third and fourth flows at Little Stony Man. As at Little Stony Man, there are two possible interpretations. One is that the material was unconsolidated sand or silt from beneath the magma that was squeezed up into the breccia as the magma flowed over the surface. The second interpretation is that the breccia in which the dikes are found was formed at the top of a flow that fractured upon cooling, creating large voids between blocks of basalt. Silt may have been deposited by wind or water around these angular blocks to form the tabular zones of siltstone you see today. At Dark Hollow Falls, I tend to favor the second interpretation, because some of the siltstone dikes are over a foot long, which would be expected if these were voids around blocky lava that were filled by sedimentary material. ■

- Distance from
  Thornton Gap
  Entrance Station
  19.8 miles

- Distance from
  Dark Hollow Falls
  Parking Area
  0.6 miles

- Next stop:
  Bearfence Mountain
  Parking Area
  5.2 miles

- Distance from
  Bearfence Mountain
  Parking Area
  5.2 miles

- Next Stop:
  Dark Hollow Falls
  Parking Area
  0.6 mile

- Distance from Swift Run
  Entrance Station
  14.3 miles

14-1 View to the
southwest from
Blackrock. The small
knobs in the center of
the photo are
composed of resistant
layers of quartzite of
the Cambrian-aged
Hampton and Erwin
formations. The
Stanley Fault is in the
valley just beyond the
knobs.

Big Meadows Area, Blackrock
Mile 51.2

At Big Meadows, mile 51.2 along Skyline Drive, turn west onto the 1-mile-long access road that leads to Big Meadows Lodge and Big Meadows Campground. Park at the south end of the lodge parking lot. From here, a short trail leads 450 feet to a lookout called Blackrock. From Blackrock you can view the approximate trace of the Stanley Fault, which cuts across the base of the west side of the Blue Ridge Mountains.

Blackrock is the eleventh metabasalt flow from the bottom of the volcanic sequence and, stratigraphically, the highest in the Big Meadows area. From this lookout you have excellent views to the west of Shenandoah Valley and the Valley and Ridge Province. The purpose of this stop is to show a major structural feature that helped form this section of the Blue Ridge Mountains and to show how topography and landforms are frequently controlled by the relative resistance to erosion of the different rock units.

First, look to the southwest (left) at the westward-trending line of ridges (see photo 14-1). Note the pronounced bumps on the ridge. This region of the park is underlain by sedimentary rocks and the bumps are composed of quartzite, which is more resistant to erosion than the surrounding sandstones and phyllites.

Next, look straight ahead toward the

STANLEY FAULT

14-2 View west from Blackrock toward the town of Stanley in the Shenandoah Valley. The approximate trace of the Stanley Fault is marked, truncating the ridge to the south (left) of town and passing just east of the town.

THORNTON GAP

STANLEY FAULT

14-3 View northwest from Blackrock. The approximate trace of the Stanley Fault is marked as it cuts northeast from the base of the Blue Ridge through Thornton Gap at mile 31.4.

town of Stanley and the ridge immediately to its left (south), which ends rather abruptly at the town (see photo 14-2). This marks the location of the Stanley Fault, which runs along the base of the mountains from the south, cutting across the truncated ridge just south of town and then continuing east. Now look to the northwest and con-

tinue to trace the fault as it passes north between the Blue Ridge Mountains and two low-lying ridges (see photo 14-3), before cutting northeast across the Blue Ridge Mountains at Thornton Gap (just out of your field of view and the field of view in photo 14-3). The trace of the fault is also shown on the accompanying topographic map. Along

Topographic map showing location of the Stanley Fault. Blackrock is marked by the red dot. Map courtesy U.S. Geological Survey.

this fault, Precambrian and early Paleo-zoic siliciclastic rocks have shifted over the early Paleozoic limestones and shales composing the valley. ■

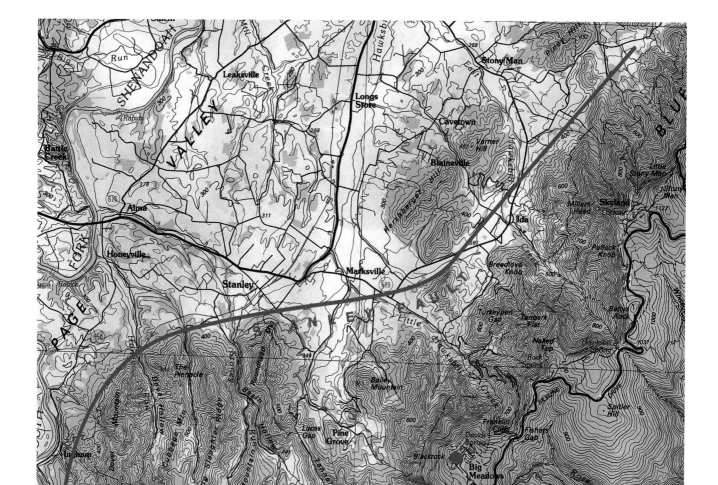

SCALE 1:100 000

1 CENTIMETER ON THE MAP REPRESENTS 1 KILOMETER ON THE GROUND
CONTOUR INTERVAL 20 METERS

Bearfence Mountain Parking Area
Mile 56.4

This is one of the best and most accessible exposures of the Swift Run Formation, which is visible at the beginning of the trail. The circuitous climb over boulders to the top of Bearfence Mountain, for those physically able, provides access to some unusual columnar-jointed basalts, as well as a magnificent panoramic view of the region. The loop trail over the top of Bearfence Mountain and back to the parking area is about 1 mile long with a 200- to 300-foot change in elevation.

SOUTHBOUND

- Distance from Thornton Gap Entrance Station
  25 miles
- Distance from Big Meadows along Skyline Drive
  5.2 miles
- Distance to Swift Run Gap Entrance Station
  9.1 miles
- Next Stop: Bacon Hollow Overlook
  12.9 miles

NORTHBOUND

- Distance from Bacon Hollow Overlook
  12.9 miles
- Next Stop: Blackrock at Big Meadows
  5.2 miles along Skyline Drive, about 1 mile on access road
- Distance from Swift Run Gap Entrance Station
  9.1 miles

Sandy phyllite, conglomerate, and quartzite of the Swift Run Formation, which underlie the Catoctin metabasalts, are exposed along the first 400 feet of the trail leading to Bearfence Mountain. At the beginning of the trail, tan, thinly bedded, sandy phyllites are exposed. Similar phyllites are better exposed in the roadcut at Hensley Hollow Overlook, mile 64.4, but the conglomerate and quartzite are better ex-posed on this trail. As the trail enters the woods, loose fragments of white-to-tan conglomerate are abundant (see photo 15-1), and a few outcrops occur to the south (right) of the trail. Fragments of conglomerate predominate over the first 200 feet of trail, then good outcrops of quartzite continue to the top of the rise (see photo 15-2), about 400 feet from the trailhead.

Because the trail over Bearfence

15-1 Swift Run Conglomerate showing fragments of white quartzite within a matrix of white sandstone. This loose rock, unattached to the bedrock, is what geologists refer to as "float", as opposed to the outcrops of solid bedrock in photo 15-2.

15-2 Outcrop of tan quartzite from the Swift Run Formation along trail to Bearfence Mountain. Notebook, for scale, is 5" x 7.5".

Mountain traverses up, over, and around numerous jagged outcrops, it is referred to as a "rock scramble" and the rocks have been given the name "Bearfence Rocks." The trail climbs for several hundred yards over metabasaltic flows and volcanic breccias of the Catoctin Formation. Early in the scramble you can see some columnar-jointed metabasalts lying on their sides, oriented between 15 and 20 degrees from horizontal with their tops pointed in a northerly direction (see photo 15-3). Another group of columns, oriented 45 degrees from horizontal with their tops pointed in a more easterly direction is only 20 to 25 feet away (see photo 15-4).

Two groups of columns with such different orientations so close to one another are problematic—how could they have formed this way? My interpretation is that the lavas that formed them originally flowed out over a relatively flat surface. As they cooled, vertical columnar joints were formed. Then the flow was eroded and undermined, and jointed segments collapsed or fell over. Finally, another flow poured over what remained of the previous flow, and the leaning columns were covered and frozen in place. This overlying flow has now been eroded, uncovering the collapsed columns.

A thick zone of volcanic breccia occurs farther to the south along the rock scramble, overlying the flow that contains the leaning columns. The contact between the metabasalt flow and the overlying breccia is clearly an irregular surface and is further evidence that the lowermost basaltic flow was partially eroded before being covered by the next flow. The metabasaltic flow and the breccia are very easy to distinguish at Bearfence Rocks. The metabasalts

15-3 Columnar metabasalt with columns lying at 15-20 degrees from horizontal and with column tops pointing N20E.

15-4 Columnar metabasalt with columns at 45-degree angle and a bearing of N75E. This set of tilted columns and the ones above, only 20-25 feet apart, are evidence of their collapse during erosion prior to the next volcanic event.

contain a metamorphic layering formed by alignment of the green mica chlorite, with a northwest orientation and gentle northeast dip (see photo 15-5). The volcanic breccia is massive, with no micaceous minerals, and therefore no layering, and abundant amygdules. The contrast between the two types of rock is shown in photo 15-6.

Amygdules are also quite evident at

15-5 Metabasalt on top of Bearfence Mountain. The layering is due to alignment of the green mica, chlorite, in the metabasalt, formed during metamorphism.

15-6 Irregular contact of volcanic breccia overlying metabasalt (contact shown with dashed line). This is clear evidence for a period of erosion of the basalts prior to the next volcanic flow.

Bearfence Mountain. As at Dark Hollow Falls, these were air vesicles within the metabasalts that filled with quartz, epidote, feldspar, and/or hematite. This type of amygdule is abundant in the volcanic breccia at Bearfence Mountain. However, there are some other amygdules in the metabasaltic flows near the top of Bearfence Mountain that are filled with chlorite. These appear as dark green or black blotches in a lighter gray metabasalt (see photo 15-7). This is one of the few locations in Shenandoah National Park where

15-7 Amygdaloidal
metabasalt with
amygdules filled
with dark green
chlorite.

15-8 View to the
north along the top
of Bearfence
Mountain. Outcrops
in the foreground are
of volcanic breccia.
The mountain in the
background on right
is Bush Mountain. The
farthest line of small
hills cutting
diagonally from the
left is part of Long
Ridge.

this type of amygdule can be observed.

After studying the rocks, be sure to
enjoy the panoramic view. ■

SOUTHBOUND

- Distance from Swift Run Gap Entrance Station
  3.8 miles

- Distance from Bearfence Mountain Parking Area
  12.9 miles

- Next Stop: Loft Mountain Overlook
  5.2 miles

NORTHBOUND

- Distance from Loft Mountain Overlook
  5.2 miles

- Distance to Swift Run Entrance Station
  3.8 miles

- Next Stop: Bearfence Mountain Parking Area
  12.9 miles

- Distance from South Entrance Stration
  35.9 miles

16-1 Smoothly weathered outcrop of Pedlar Formation with gneissic banding and spaced fractures called "joints." Gneissic banding developed during metamorphism when the rock was heated and folded. The jointing cuts the banding and originated at a later time, perhaps during a late stage of folding or when rocks in the area were gradually uplifted and stresses were relieved.

Bacon Hollow Overlook
Mile 69.3

If you are headed south on Skyline Drive, this is your last chance to see the Pedlar Formation before Skyline Drive moves on to younger, overlying strata.

At this location the Pedlar is light in color, containing mostly feldspar and quartz and minor amounts of the dark minerals hornblende and pyroxene, but no visible garnet. The quartz grains are gathered into distinct, thin layers (see photo 16-2), producing the banding called gneissic texture. This banding is characteristic of strongly folded and metamorphosed rocks commonly found in the older portions of mountain chains.

Looking down into Bacon Hollow, where the soil is exposed from recent housing developments and all-terrain vehicle tracks, you can see pronounced red soils developed from weathering of the Pedlar Formation. Red, or *lateritic,* soils contain the iron oxide mineral hematite. They commonly develop in southern climates where iron-bearing silicate minerals decompose by reacting with oxygen in the atmosphere and by dissolving in water. Red soils

◀ ▪▪▪

16-2 Banded gneiss of the Pedlar Formation. Layering is formed by bands of quartz grains and the dark iron-bearing minerals pyroxene and hornblende, with intervening zones of cream-colored feldspar. Rust color is due to oxidation of the iron in the dark minerals.

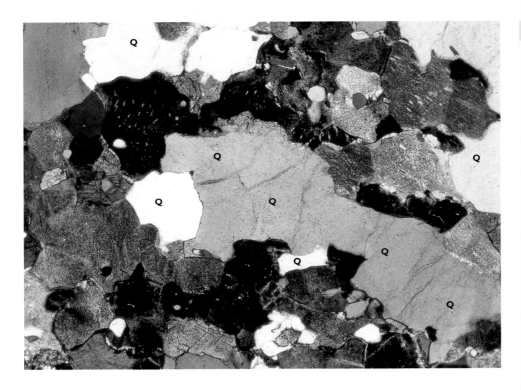

◀ ▪▪▪

16-3 Microscope view of a thin section of Pedlar Formation using crossed polarized light. Note the group of quartz grains across the center of the photo. Clumps of such grains forming irregular bands give the rock its "gneissic" texture. Field of view is 3 mm. Q = quartz; most of the remaining minerals are feldspar.

develop where the Pedlar Formation's iron-bearing silicates, pyroxene and hornblende, have decomposed due to weathering.

Flattop Mountain, to the south, part of which is outside of the park boundary, is visible to the south and is capped by Catoctin metabasalts. ∎

16-4 Bacon Hollow, one of many valleys carved into the mountains. Areas of housing development and forest clearing show the red color of the underlying lateritic soils. The weathering and oxidation of iron silicate minerals in the Pedlar Formation has produced the reddish iron oxide mineral hematite (Fe_2O_3).

Loft Mountain Overlook
Mile 74.5

The outcrop of Catoctin metabasalt at this overlook belongs to one of the volcanic flow units. It contains several features characteristic of the tops of these lava flows, including cream-colored amygdules, irregular zones of purple oxidized metabasalt, striations called slickenlines, *and pods of pistachio-to olive-green epidote. Climb on this outcrop and look for each of these features.*

As discussed at Dark Hollow Falls (see Stop 13, page 51), the presence of amygdules is indicative of the degassing of the upper surface of a still-molten lava flow. The loss of steam and carbon dioxide from the magma left bubbles and air pockets, which formed a frothy surface layer called *scoria*. After the magma solidified, the pockets were filled by quartz, epidote, feldspar, and/or hematite to form amygdules.

Purplish zones in the metabasalt are interpreted as parts of lava flows that were exposed to the atmosphere for hundreds or thousands of years before being buried by another lava flow. During a time of extensive weathering and oxidation, the iron in the flow, commonly 13 to 14 percent of the total mass, was oxidized to hematite, thus creating the purple color (see photo 17-2).

SOUTHBOUND

- Distance from Swift Run Entrance Station
  9 miles
- Distance from Bacon Hollow Overlook
  5.2 miles
- Next stop: Rocky Top Overlook
  3.6 miles

NORTHBOUND

- Distance from Rocky Top Overlook
  3.6 miles
- Next Stop: Bacon Hollow Overlook
  5.2 miles
- Distance from South Entrance Station
  30.7 miles

17-1 Outcrop at Loft Mountain Overlook consisting of the top portion of a metabasaltic flow unit. The following three photographs are close-ups of features seen in this outcrop.

17-2   Purple oxidized portion of the metabasaltic flow. Grasshopper on lower left for scale.

17-3 Slickenlines aligned parallel to the pen indicate the direction of relative movement.

Slickenlines are parallel lines or narrowly spaced, shallow grooves formed by the movement of one mass of solid rock over another (see photo 17-3). These etched and scratched surfaces are commonly found on faults or in zones where the rocks were stressed. The contact surface of one lava flow with another would have been a natural zone of weakness in the pile of lava flows that

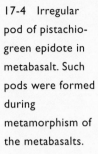

17-4 Irregular pod of pistachio-green epidote in metabasalt. Such pods were formed during metamorphism of the metabasalts.

composes the Catoctin Formation. Thus, if stresses were applied, slippage would likely have occurred along the contact zone. Folding, faulting, and uplift of the rocks during formation of the Appalachian Mountains provided the stresses that led to the localized slippage and shearing to produce the slickenlines.

Oval pods of bright, pistachio- to olive-green rock within the metabasalt are composed of the mineral epidote (see photo 17-4). Laboratory work by Jack Reed and Ben Morgan of the U.S. Geological Survey showed that these epidote pods were formed by cation exchange during metamorphism as hot watery fluids slowly seeped through the rocks. The fluids transported calcium from one portion of the basalt and deposited it in localized areas to form the epidote pods. At the same time, magnesium and sodium were transported from the emerging epidote pod into the surrounding basalt. The pods are frequently cut by veins of white-to-clear quartzite that were also produced at this time. The epidote pods and quartz veins are characteristic of the Catoctin Formation and are particularly well displayed here. Formation of the epidote pods occurred during metamorphism associated with the folding of the rocks during continental collision, about 450 million years ago.

Loft Mountain, to the south from the overlook, is also underlain by Catoctin metabasalts. ■

SOUTHBOUND

- Distance from Swift Run Gap Entrance Station
  12.6 miles

- Distance from Loft Mountain Overlook
  3.6 miles

- Next Stop: Doyles River Overlook
  3.8 miles

NORTHBOUND

- Distance from Doyles River Overlook
  3.8 miles

- Next stop: Loft Mountain Overlook
  3.6 miles

- Distance from South Entrance Station
  27.1 miles

18-1 Talus slopes and cliffs of resistant Erwin quartzite produce the light-colored treeless areas on Brown Mountain. These highly resistant quartzites contain fossilized Skolithos tubes.

Rockytop Overlook
Mile 78.1

GEOLOGY ·18· STOP

On the slopes of Rockytop Mountain to the southwest and Brown Mountain to the west, look for cliffs and talus slopes of quartzite of the Erwin Formation.

The Erwin Formation is the youngest and stratigraphically, the highest siliciclastic unit of the sedimentary sequence overlying the Catoctin volcanics in Shenandoah National Park. This overlook offers views of the protruding cliffs formed by these very resistant quartzite beds. Sandstones and phyllites of the Hampton Formation underlie the quartzite. They erode more easily than the quartzite cliffs, which results in undermining, or loss of support. The undermined quartzite breaks into fragments that roll downhill to form the talus slopes, piles of eroded rock.

Conveniently, the rock wall at this overlook is composed predominantly of the Erwin Formation, so it can be readily viewed. This white, medium- to coarse-grained quartzite was once beach sand that was deposited by the sea that invaded the area after eruption of the Catoctin volcanics approximately 570 million years ago. Many of the rocks on the wall contain Skolithos tubes, fossilized vertical tubes that represent the burrowing of small worms into the beach sand (see photo 18-3). The holes quickly filled with sand, but the filled-in burrows have a slightly different

18-2 Outcrops of quartzite of the Erwin Formation at south end of Brown Mountain. The milky, vitreous-lustered quartz makes for a hard, brittle rock. Under high pressure and great depth of burial, however, the layers of quartzite will bend into large folds.

18-3 Skolithos tubes in a block of quartzite in the rock wall at Rockytop Overlook. The tubes are the 2-to-3-mm-wide, vertical, pencil-like markings in the rock; they are the fossilized burrows of Cambrian-aged worms in a beach sand.

texture than the surrounding quartzite, so they stand out in relief. The fossil traces are indicative of the Cambrian period, about 500 million years ago (see page 78 for further discussion of Skolithos tubes).

Although not apparent from this overlook, the layers of quartzite at Brown Mountain have been bent into a large fold. This fold can be seen from the south. (See photo 18-4, taken from the Rockytop Trail.) The only way to form folds in such massive quartzite is to bury the rocks under several thousand feet of rock and sediments so that it heats up, and then apply a lateral

18-4 Folded quartzite (note the curving layers of rock) at the south end of Brown Mountain. Generally, rocks as brittle as quartzite are deformed by faulting instead of folding. Photo taken from Rockytop Trail.

18-5 Sandstone ledges of Hampton Formation along road adjacent to Rockytop Overlook. Note the layering of the sandstone dipping gently to the right (east). Sandstone differs from the quartzite layers because small amounts of clay and mud prevented the sand grains from being as tightly compacted as the sand grains in a quartzite, when the material was compacted into rock.

stress, such as would occur during continental collision. This is what geologists believe happened at Brown Mountain. The area was squeezed during one of the mountain building events that led to the uplift of the Appalachian Mountains. The overlying sedimentary rocks have long since eroded, leaving the exposed, folded layers of quartzite.

At the roadcut behind the overlook, notice the massive ledges of tan-to-gray sandstone of the Hampton Formation (see photo 18-5). ■

Doyles River Overlook

Mile 81.9

Take a short walk on the Appalachian Trail to see exposures of quartzite and phyllite of the Hampton Formation. Do not confuse the overlook with Doyles River Parking Area at mile 81.0.

SOUTHBOUND

- Distance from Swift Run Gap Entrance Station
  16.4 miles
- Distance from Rockytop Overlook
  3.8 miles
- Next Stop: Blackrock Parking Area
  2.6 miles

NORTHBOUND

- Distance from Blackrock Parking Area
  2.6 miles
- Next Stop: Rockytop Overlook
  3.8 miles
- Distance from South Entrance Station
  23.3 miles

This overlook features boulders of coarse quartzite of the Hampton Formation. Look closely at some of the rocks and you may note the presence of several layers, a few centimeters thick, that vary in grain size. The base of each layer begins with coarse quartz sand and pebbles up to a quarter inch in diameter that grade upwardly to finer-grained, sand-sized quartz fragments. The layers are repeated throughout the rocks, a feature called graded bedding. Each layer was probably deposited at the bottom of a shallow sea during a single event. A surge of sediment-bearing water, perhaps churned by the turbulence of a storm, passed into deeper and quieter water; the heavier particles settled out first, followed by the finer-grained particles.

Take a short walk along the Appalachian Trail heading south; you will

19-1 Ledges of quartzite of the Hampton Formation along Appalachian Trail about 900 feet south of Doyles River Overlook. Note the nearly horizontal layering, which is the original bedding of the material.

19-2 Ledges of quartzite of the Hampton Formation about 1,100 feet south of Doyles River Overlook. Note gap near bottom where a thin zone of phyllite has eroded. The bottom of the overhang is an irregular surface and represents the contact of a coarse-grained sand layer on a now eroded thin layer of mud.

19-3 Contact between thinly layered phyllite and overlying quartzite of the Hampton Formation. Note the nearly horizontal layering of the quartzite, which is bedding, and the inclined layering of the phyllite, which is due to metamorphic alignment of micas. Notebook is 5" wide.

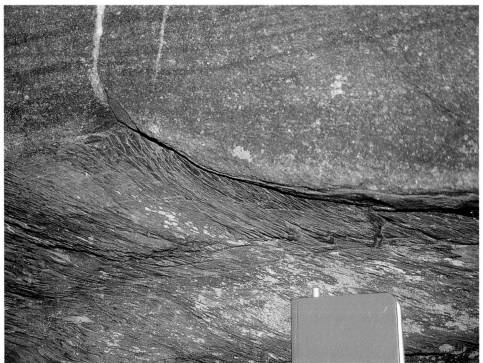

walk along the top of a small ridge underlain by this resistant quartzite. The trail stays on the ridge for about 700 feet, with float and a few outcrops of quartzite visible, before gently declining to the east to the base of a ledge (see photo 19-1). The trail runs along the base of the ledge, which is composed of quartzite with thin beds of gray sandstone and greenish gray phyllite.

Near the south end of the ledge, there is an excellent example of the contrasting behavior of quartzite and phyllite with respect to erosion and the different ways that sand and mud compact to form rock. The phyllite layer, only a few inches thick, eroded faster than the quartzite, leaving an indentation, or gap (see photo 19-2). The base of the quartzite layer extends over the gap, and that base is a gently rolling irregular surface (see photo 19-3).

Originally, the quartzite and phyllite were sand and mud, respectively. As the weight of the sand increased on top of the mud, the mud was squeezed into the irregular rolling pattern that we now see at the contact between the two rock layers. Also, quite strikingly, the thinly layered phyllite displays the dominant orientation of layering throughout this area, trending north-northeast and dipping to the southeast, while the quartzite is massive and displays no layering (see photo 19-3).

The layering in the phyllite can easily be mistaken for the orientation of the original sedimentary beds, but it is not. The original bedding plane between the quartzite and phyllite, which can be seen at the contact between the two and, upon close inspection, in the phyllite, is nearly horizontal, while the layering in the phyllite is steeply inclined. Likewise, the layering is not what geologists call *crossbedding*, which is bedding formed at an angle to the horizontal, commonly found on sand dunes along shore or in the desert or on the flanks of small mounds on the ocean floor. Instead, the layering is caused by metamorphism of the rocks and the alignment of flat-lying minerals, such as micas, within the phyllite. Its contrast with the overlying massive quartzite, which contains no mica or other flat-lying minerals that would reflect the metamorphic layering, is quite remarkable. ■

SOUTHBOUND

- Distance from Swift Run Gap Entrance Station
  19 miles

- Distance from Doyles River Overlook
  2.6 miles

- Next stop: Horsehead Mountain Overlook
  4.2 miles

NORTHBOUND

- Distance from Horsehead Mountain Overlook
  4.2 miles

- Next stop: Doyles River Overlook
  2.6 miles

- Distance from South Entrance Stration
  20.7 miles

20-1 Rubble (talus) and ledges of quartzite at Blackrock. The quartzite is light colored, almost white, but is covered by rock tripe, a lichen, which gives it a darker color.

Blackrock Parking Area
Mile 84.5

GEOLOGY · 20 · STOP

Take an easy 1-mile loop trail to a panoramic view from Blackrock, a well-exposed knoll of quartzite, which is part of the Hampton Formation. This is not to be confused with Blackrock at Big Meadows, Stop 14, which is composed of metabasalt.

This is the same massive quartzite unit of the Hampton Formation as seen at Stop 19, but with a spectacular panoramic view of the mountain scenery. Blackrock consists of a large area of ledges and quartzite rubble, the products of weathering. This is the local high point, so the talus falls in all directions, away from this quartzite knoll. The rock is light gray to tan in color, but it is called Blackrock because of the presence of rock tripe, a black lichen that grows on the rock. More geologic features, such as the interbedded phyllite and contrast in layering, can be seen in the outcrops at Doyles River Overlook, Stop 19, but the short, gentle hike to Blackrock is included here because of the magnificent scenery in all directions and a good rock slide to climb on. ■

Horsehead Mountain Overlook

Mile 88.7

GEOLOGY **·21·** STOP

Acid precipitation is decimating Shenandoah's aquatic habitat. At this stop we will discuss the effects of acid precipitation on Shenandoah's waters and the acid-neutralizing capacity of the various types of bedrock in the park.

SOUTHBOUND

- Distance from Swift Run Entrance Station
  23.2 miles
- Distance from Blackrock Parking Area
  4.2 miles
- Next stop: Riprap Parking Area
  1.3 miles

NORTHBOUND

- Distance from Riprap Parking Area
  1.3 miles
- Next stop: Blackrock Parking Area
  4.2 miles
- Distance from South Entrance Station
  16.5 miles

In the valley before you is the watershed of Paine Run, one of many small streams whose headwaters lie within Shenandoah National Park. These streams once provided good habitat for more than a dozen species of native fish, including brook trout, rosyside and blacknosed dace, fantail darters, and sculpins. This is no longer the case. Paine Run is a beautiful stream, but its water has become acidified and toxic to all but the most resilient organisms.

The water in Paine Run has a pH (the measure of acidity or alkalinity) that has recently ranged from 5.0 to 6.2, which is well below the neutral value of 7. (Waters below 7 are acidic, those above are basic.) The watershed is underlain by siliciclastic rocks, which cannot neutralize the acidity of water that flows over them. Of the many fish species that once called this stream home, only a few brook trout and blacknosed dace survive here today. Streams such as Paine Run contain diminished numbers of crayfish, frogs, and salamanders that crawl on its rocky banks; in the

21-1 Paine Run watershed from Horsehead Mountain Overlook. The stream runs through the small valley in the center of the photo.

spring there are few hatches of mayflies and stoneflies. Paine Run is a beautiful stream to look at, but like many streams in Shenandoah, its web of life has been severely damaged.

The causes of the acid precipitation in Shenandoah's streams are sulfates and nitrates contained in the air and water vapor that drift over the park, predominantly from Midwestern sources. Look west from the overlook. How far can you see? Is Massanutten Mountain clearly visible or is it obscured by a brownish gray haze? How about the Allegheny Mountains beyond Massanutten? The haze of pollution that commonly obscures the views from Shenandoah National Park is a visible reminder of the major environmental problem facing the park—the acidifi-

cation of its waters.

The problem of acid precipitation in the park has been studied since the late 1970s by an interdisciplinary group of atmospheric chemists, geochemists, hydrologists, and ecologists from the University of Virginia, in cooperation with scientists from the Park Service and the U.S. Geological Survey. From their data and analyses, there emerges a very frightening scenario for Shenandoah's watersheds and streams (see References for citations).

The Air Resource Division of the National Park Service estimates that every acre in the park receives 25 pounds of sulfates each year as an accompaniment to rain, fog, and snow. The table below lists the sources of these pollutants.

Sources of airborne sulfates reaching Shenandoah National Park every year.

Pittsburgh-Cleveland area	28.8%
Piedmont-Northern Tennessee	17.8
Columbus-Dayton-Cincinnati	8.5
Chicago	6.7
Central Michigan	4.9
Toledo	4.8
Austin-Houston	4.0
Western New York	3.6
Mid Ohio River	3.2
Other	17.7

Source: Air Reserve Division of the National Park Service.

21-2 Paine Run with University of Virginia scientist Ishi Buffam collecting a water sample for analysis. The rock along the shore is quartzite, which has almost no acid-neutralizing capacity.

A comparison of the sulfate content of precipitation at 13 U.S. national parks from 1991 to 1995 was compiled by Rick Webb at the University of Virginia (see figure 12, page 77). The chart shows that this is clearly an Eastern problem, with Shenandoah and Smoky Mountain national parks receiving three to ten times the amount of sulfate deposition as such popular western parks as Yosemite, Yellowstone, Rocky Mountain, and the Grand Canyon. This clearly reflects the Midwestern source areas for the sulfates and the prevailing winds.

Normal precipitation, unaffected by pollutants, has a pH of about 5.5 and can be described as slightly acidic. However, the pH of precipitation falling in Shenandoah National Park averages about 4.2. Each unit on the pH scale represents a tenfold change in acidity,

so a pH of 4.2 is more than ten times more acidic than normal precipitation. Some of this precipitation will flow over the land directly into streams. The remainder will soak into the soil, slowly seeping through it, eventually entering the streams.

Soil is composed of weathered rock fragments, clay-sized grains and larger, mixed with organic material. The rock fragments are weathered from the underlying bedrock, so the chemistry of the soil reflects the composition of the bedrock. If the soil contains positively charged ions, called cations, such as calcium or magnesium, the acids can be neutralized. The ability of a soil to break down acids is called the *acid-neutralizing capacity* of the soil. The acid-neutralizing capacity of a watershed is controlled by the bedrock and its ability to neutralize acids. If a watershed has a high acid-neutralizing capacity, acids will be neutralized as the water moves through the soil, and the stream will have a healthy pH, in the range of 6.5 to 8. Limestone is the best rock for neutralizing acids, but the limestones in northern Virginia are found in the Shenandoah Valley, not within Shenandoah National Park.

In Shenandoah National Park there are three main types of bedrock, classified by origin and chemical composition: the volcanic rocks of the Catoctin Formation, the gneisses and granites of the Pedlar and Old Rag formations, and the siliciclastic quartz-bearing sedimentary rocks of the Weverton, Hampton, and Erwin formations (see map on page 2). Weathering in each area of bedrock has produced soils of different chemical content and therefore different acid-neutralizing capacities. The Catoctin Formation is relatively effective in neutralizing acids because the breakdown of the minerals pyroxene and feldspar readily provide ions of calcium, magnesium, and iron. (See geochemical data for the rocks in Appendix 2.) So, watersheds underlain predominantly by Catoctin metabasalts are fairly healthy, with pH levels in their streams ranging from 6.6 to 7.3. These streams typically support up to 15 different species of fish.

The Pedlar and Old Rag formations provide some cations to stream waters by the breakdown of feldspars and lesser amounts of pyroxene or amphibole that are present. Basaltic dikes that cut across the Pedlar and Old Rag formations also contribute cations, but the overall cation supply is inadequate to completely neutralize the acids. Streams in these watersheds have a pH in the range of 6.0 to 7.1 and commonly support five to eight species of fish.

The siliciclastic rocks—the conglomerates, sandstones, phyllites, and quartzites of the Weverton, Hampton and Erwin formations—are composed chiefly of quartz (SiO_2) and have very little acid-neutralizing capabilities. Watersheds underlain by these rocks have a pH in the range of 4.8 to 6.2, and many streams in these areas are unable to support more than two species of fish: brook trout, which are remarkably resilient, and blacknosed dace. Other species that once flourished in these sparkling waters have died off. Also affected are those organisms at the base of the food chain, such as tadpoles, mayflies, and caddis. The water may become intolerable even for bacteria, so that organic material such as leaves and pine needles fail to decompose and remain at the bottom of streams—clear water is *not* unpolluted water.

Acidified water affects fish populations in several ways. Sublethal effects include deformities of gills or fins, decreased size, and contorted or abnormal body shapes. Reproductive failures occur when females lay fewer eggs and many of those fail to hatch. A 90 to 95 percent rate of hatching success may drop to a 30 percent success rate in acidified water, and many of those hatchlings will contain deformities. The final result is a loss of species. No young are born, either because females were no longer able to lay eggs or the eggs failed to hatch, and the older fish slowly die off.

Is there hope on the horizon? The situation is bleak. The Clean Air Act is supposed to decrease emissions of sulfates at their sources outside the park. Sulfate reduction will lead to an improvement in air quality, but much of the damage to the watershed is not reversible. Studies by University of Virginia scientists show that much of the acid-neutralizing capacity of the soils overlying siliciclastic bedrock has been used up, and that little potential remains for neutralizing more acid precipitation. Sadly, even if air quality is improved and precipitation becomes less acidic, those portions of the park underlain by siliciclastic rock lack the proper soil chemistry to bring about a recovery.

To see the effects of acid precipitation in the park, spend some time examining the aquatic organisms in Paine Run or some other stream that is underlain by siliciclastic rock. Then examine the diversity of species living in a stream underlain by metabasalt, such as the Rose River. The rocks really do make a tremendous difference in their contribution to soil and water chemistry, and hence to the biodiversity in the different watersheds. ∎

Figure 12

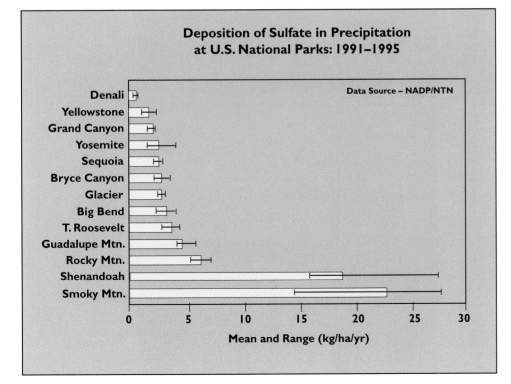

Deposition of Sulfate in Precipitation at U.S. National Parks: 1991–1995

Graph showing deposition of sulfate in precipitation at U.S. national parks. Graph constructed by Rick Webb of the University of Virginia and used with his permission.

SOUTHBOUND

■ Distance from Swift Run Gap Entrance Station
24.5 miles

■ Distance from Horsehead Mountain Overlook
1.3 miles

■ Next stop: Sawmill Ridge Overlook
5.9 miles

NORTHBOUND

■ Distance from Sawmill Ridge Overlook
5.9 miles

■ Next stop: Horsehead Mountain Overlook
1.3 miles

■ Distance from South Entrance Station
15.2 miles

22-1 Ledges of cream-colored quartzite of the Erwin Formation at Calvary Rocks. These rocks represent beach sand, made of quartz, that has become lithified into a hard quartzite.

Riprap Parking Area

Mile 90

Take a leisurely hike to Calvary Rocks to see quartzite of the Erwin Formation and the fossilized burrows of the Cambrian-aged worm Skolithos. The hike is an easy 1.2 miles to Calvary Rocks, so the round trip is 2.4 miles, not 3.4 miles as shown on the trailhead sign.

This is the most accessible location for viewing quartzite outcrops of the Erwin Formation, the highest unit of the siliciclastic sedimentary rocks that overlie the Catoctin metabasalts (see photo 22-1). Preserved within the quartzites are abundant Skolithos tubes, the fossilized burrows of Cambrian-aged worms. The worms dug and lived in vertical burrows in what was once loose beach sand composed of quartz grains. The burrows are readily visible on the fractured or weathered surfaces cut perpendicular to bedding (see photo 22-2). On surfaces that are parallel to the bedding, look for cross-sectional

22-2 A vertical cross-section showing Skolithos tubes dug in sand (now quartzite) by worms during the Cambrian time period. The vertical orientation of the tubes has changed little since that time.

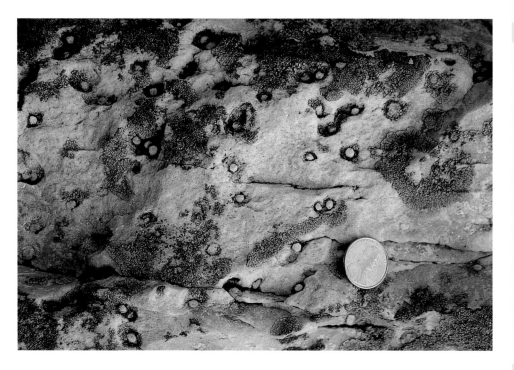

22-3 Looking down on the quartzite and the small circles that are the ends of Skolithos tubes. A penny is shown for scale.

views of the burrows—small circles that are 2 to 3 mm in diameter (see photo 22-3).

The worms that dug these burrows are believed to have lived during the early to middle Cambrian time period, approximately 544 to 525 million years ago. The presence of these tubes was, for years, the only clue to the age of the siliciclastic rocks and underlying volcanics. If there was a gap of tens of millions of years between eruption of the Catoctin volcanics and deposition of the overlying sediments, then you would expect major portions of the Catoctin volcanic rocks to have been eroded before the overlying sedimentary rocks were deposited. This is not the case.

Although there is some evidence of local erosion of volcanic flows (see discussion in Stop 15, page 56), in general the flows are conformably overlain by conglomerates and sandstones of the Weverton Formation, meaning there had been no apparent break in time between deposition of the volcanic and sedimentary rocks. Likewise, sedimentary rocks of the Weverton Formation are conformably overlain by sandstones of the Hampton Formation, followed by quartzites of the Erwin Formation. So if the age of the quartzites of the Erwin Formation is Cambrian, the underlying volcanics cannot be much older; most likely they date to the late Precambrian period.

The first radiometric age published for the Catoctin volcanics was 820 million years. This was too old, based on the fossil and stratigraphic evidence, and even the authors of that paper admitted the age did not agree with the fossil evidence. A subsequent published age of 490 million years, an Ordovician age, was equally suspect because of the Cambrian-aged Skolithos tubes. More recent radiometrically determined ages of 570 (my determination) and 565 (U.S. Geological Survey) million years are compatible with the fossil evidence. For further discussion of the methods of dating rocks, see Appendix 3. ■

Sawmill Ridge Overlook
Mile 95.9

This is the best location along Skyline Drive to see the sandstones and siltstones of the Weverton Formation, the sedimentary rocks that immediately overlie the Catoctin metabasalts.

The outcrops along the roadcut are composed of massive and thinly bedded, greenish gray sandstone and siltstone with pod-shaped layers of conglomerate. These rocks have been assigned to the Weverton Formation, the lowest unit of the sedimentary rocks that overlie the Catoctin volcanics. This is the best location along Skyline Drive to see a good exposure of the Weverton Formation. The Weverton and Hampton formations (see Stops 19 and 20) are very similar clastic sedimentary rocks in that both contain a mixture of conglomerates, sandstones, and phyllites, but conglomerates predominate in the Weverton, while quartzites and sandstones predominate in the Hampton Formation. The Weverton underlies the Hampton and is therefore older. ■

SOUTHBOUND

- Distance from Swift Run Gap Entrance Station
  30.4 miles
- Distance from Riprap Parking Area
  5.9 miles
- Distance to South Entrance Station
  9.3 miles

NORTHBOUND

- Distance from South Entrance Station
  9.3 miles
- Next stop: Riprap Parking Area
  5.9 miles

23-1
Conglomeratic sandstone of the Weverton Formation along the roadside at Sawmill Ridge Overlook. The layering dipping gently to the right is the original bedding of the rock.

23-2 Conglomeratic sandstone of the Weverton Formation. The mottled appearance of the rock below the knife is due to coarse, rounded grains and pebbles of quartz, up to 1/4 inch in diameter, in a sandstone matrix.

Appendix 1: Old Rag Mountain

Take a 7.1-mile loop hike to see the Old Rag Granite and basaltic dikes that cut across it. This is perhaps the most spectacular and enjoyable hike in the entire park, but parts are rather strenuous. The rewards, both geological and scenic, are well worth the day spent climbing this mountain.

Old Rag Mountain is an isolated peak east of the main ridge of the Blue Ridge Mountains; it is not along Skyline Drive. There are good views of Old Rag at Pinnacles Overlook, mile 35.1; Thorofare Mountain Overlook, mile 40.6; and Old Rag Overlook, mile 46.5. To get to Old Rag, leave the park at Thornton Gap, heading east on U.S. Highway 211. In approximately 7 miles, at Sperryville, turn right on US 522. Drive south on US 522 for 0.9 mile to Virginia Highway 231, and turn right (south). Continue on VA 231 for approximately 8.4 miles until the highway crosses the Hughes River. Immediately turn right and continue on this road, making sure to stay just to the left of Hughes River. Old Rag Mountain rises to your left (see photo A1-1). The road changes its number several times, but ignore that and stay with the river to the trailhead, at approximately 4.3 miles from the turn off VA 231. If the trailhead parking lot is full, there is an overflow parking lot at 3.5 miles.

A1-1 Southerly view of Old Rag Mountain from the Hughes River valley.

A1-2 Old Rag Granite. Blue-gray minerals are quartz, whitish grains are feldspar. Note faint alignment of minerals from upper left to lower right. This is gneissic texture, formed during metamorphism of the rocks.

VERTICAL FRACTURES

A1-3 View of Old Rag Mountain from Thorofare Mountain Overlook, mile 40.6 along Skyline Drive. Arrows point to some of the vertical, northwest-trending fractures cutting the mountain that can be seen in the late afternoon light. Some of these fractures provided conduits for Catoctin magmas.

The trail to the top is 2.6 miles long. The first 2 miles or so are in the woods, but then the trail emerges onto the rocky ridge and becomes an entertaining rock scramble around, over, and under boulders and through several narrow crevasses.

Old Rag Mountain is composed of granite, which in turn is composed primarily of bluish gray quartz and cream-colored feldspar. The rock is very coarse grained with some feldspar crystals up to 1 inch in diameter (see photo A1-2). A series of north-trending fractures cut

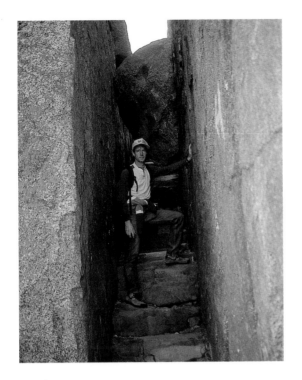

A1-4 Crevasse formed by erosion of basaltic dike cutting through Old Rag Granite. Basaltic lava once filled the crevasse where the author stands. Note step-like appearance of floor of crevasse, which consists of basalt, while vertical walls are of the Old Rag Granite.

A1-5 Erosional remnant, resembling a large Idaho potato, near the top of Old Rag Mountain. Such features would be interpreted as glacial erratics in New England mountains, but Shenandoah National Park has not been glaciated. The granite in the boulder is the same as the bedrock beneath it, so it has not been transported but was eroded in place.

across the mountain and are perhaps best viewed from a distance, such as at Thorofare Mountain Overlook in the late afternoon (see photo A1-3).

These fractures provided conduits for the Catoctin basaltic magmas to intrude the granite about 570 million years ago. Since that time, the volcanic and sedimentary rocks deposited over the Old Rag Granite have eroded, exposing the granite and the feeder dikes. Since the granite is more resistant to erosion than the highly fractured basaltic material in the dikes, the dikes

have weathered faster, leaving narrow, steep-walled passageways through the rock.

The Ridge Trail that climbs to the top of Old Rag Mountain uses two of these crevasses as passageways (see photo A1-4). These eroded dikes provide a superb example of how different types of rock weather at different rates. When you pass through the crevasses, note the rock beneath your feet. If it is fine-grained, dark green to black basalt with crosscutting fractures that make it look like stairs, you are walking on what is left of a basaltic dike. The Pedlar Formation, through which most of the feeder dikes intrude elsewhere in the park, weathers at the same rate as the Catoctin basaltic dikes, so we do not see these deep crevasses elsewhere in the park.

From the top of the mountain at the south end of the ridge, the trail descends into the woods and winds its way down the mountain. It eventually connects with the Weakley Fire Road leading back to the parking lot, a total hiking distance of 7.1 miles. ∎

Appendix 2

Many park visitors are curious about the composition of the Catoctin metabasalts. An older version of the nature guide to Stony Man Nature Trail claimed that the green color of the rocks was due to the presence of copper minerals. This is not true. The green color is due to the presence of the green, micaceous mineral chlorite and the pistachio-green mineral epidote, both of which are common minerals produced by the alteration of rocks that are rich in iron and magnesium by hot water during periods of metamorphism.

Although copper was mined at a few locations in the park, in most rock samples it is present in concentrations of less than two hundred parts per million. So, for the benefit of those who are interested, I offer chemical analysis of three samples of the Catoctin metabasalts and one sample of granodiorite from the Pedlar Formation. The first two samples were taken from the first and second flows at Hawksbill Mountain; the third is from one of the feeder dikes that cuts Old Rag Mountain. The fourth sample consists of rock from the Pedlar Formation near mile 37.7 along Skyline Drive. The oxides are given in weight percent, and the trace elements in parts per million. ∎

Chemical analysis of oxides in four samples (in percent by weight)

Oxides	Flow #1 Hawksbill Mountain	Flow #2 Hawksbill Mountain	Dike at Old Rag Mountain	Pedlar Formation
SiO_2	48.55	49.72	49.25	61.74
TiO_2	2.01	2.04	2.63	1.32
Al_2O_3	14.49	13.90	13.90	16.30
Fe_2O_3	13.48	13.10	14.67	8.33
MnO	0.22	0.23	0.24	0.12
MgO	5.28	8.04	5.92	1.63
CaO	9.60	7.64	10.25	4.48
Na_2O	3.93	2.51	2.32	2.80
K_2O	0.33	0.26	0.81	2.86
P_2O_5	0.26	0.25	0.34	0.30
H_2O	2.87	2.88	0.64	0.11
Total	101.02	100.57	100.97	99.99

Chemical analysis of trace elements in four samples (in parts per million)

Trace Elements	Flow #1 Hawksbill Mountain	Flow #2 Hawksbill Mountain	Dike at Old Rag Mountain	Pedlar Formation
Rb	9	7	48	88
Sr	242	86	282	208
Ba	119	143	160	720
Sc	44	44	35	22
Cr	139	135	64	32
Co	54	59	77	25
Ni	96	90	39	7
Cu	135	143	149	7
Zn	114	154	102	134
Zr	159	153	188	426
Nb	14	15	21	21
La	16	23	17	71
Ce	35	41	36	116
Nd	18.9	19.9	29	59
Sm	5.44	5.74	7.16	11.65
Eu	1.38	1.45	2.07	1.98
Tb	0.95	0.99	1.12	1.31
Yb	3.11	3.08	3.03	3.54
Lu	0.42	0.45	0.43	0.53
Hf	4.16	4.21	5.01	12.90
Ta	1.24	1.21	3.07	2.13
Th	1.96	2.05	1.77	1.00

Appendix 3

HOW DO WE KNOW THE AGES OF THE VOLCANIC ROCKS?

Before the days of dating rocks by radioactive methods, geologists had worked out a time scale using fossils. By identifying fossils and the succession of sedimentary rocks that contained them, geologists could determine the *relative* time period in which a certain organism appeared, when it flourished, and when it became extinct. If a rock formation of unknown age in another place was found to contain that fossil, then the formation could be placed on the geologic time scale: late Cambrian, early Permian, etc.

There are no known fossils in the Catoctin Formation because it consists primarily of volcanic rocks. However, fossils are found in the overlying metasedimentary rocks. The third rock unit above the Catoctin, the Erwin Formation, contains the fossilized burrows of a prehistoric worm known as Skolithos (see Stop 22, page 78). This particular fossil is from the early to middle Cambrian time period.

The most recent absolute age estimates for the Cambrian time period place it between 544 to 500 million years ago, so early to middle Cambrian would range from approximately 544 to 525 million years ago. The Catoctin Formation is three formations below the Erwin Formation, with the Hampton and Weverton formations in between, so it must be older. But how much older—10, 50, or even 100 million years?

As discussed at Stop 22, deposition of sediments on top of the basaltic lava flows was a relatively continuous process, and there is little to suggest that the lavas eroded before the sediments were deposited. If evidence existed that the Catoctin Formation had eroded significantly before deposition of the overlying siliciclastic rock units, then a case could be made that a stratigraphic break had occurred between the Catoctin and the overlying Weverton, Hampton and Erwin formations—a break that could represent a time gap of many tens of millions of years. No evidence exists for such a break. The fossil and stratigraphic evidence, therefore, tells us that the Catoctin Formation was deposited during late Precambrian to early Cambrian times, probably between 600 and 540 million years ago.

The first radiometric age obtained for the Catoctin volcanic lavas was 820 million years. However, even the authors of the paper reporting that age admitted that their findings did not seem to agree with evidence obtained by stratigraphic and fossil means. Their study used the decay of uranium (U) to lead (Pb), a method based on the known decay rate of two uranium *isotopes* (atoms containing an abnormal number of neutrons in the nucleus), ^{238}U, which decays to ^{206}Pb, and ^{235}U, which decays to ^{207}Pb. The superscript is the atomic weight of the element, that is, the sum of the protons and neutrons in the nucleus.

These isotopes are found in tiny grains of a mineral called *zircon*, which occurs in many igneous rocks. Zircon does not incorporate lead into its crystal structure when it crystallizes from a magma, but it does incorporate a small amount of uranium atoms, which replace some of the atoms of zirconium. Any ^{206}Pb and ^{207}Pb found in the mineral must originate strictly from the radioactive decay of the uranium. To calculate the age of the rocks, three groups of numbers are needed: (1) the relative abundance of the uranium isotopes that have not yet decayed, (2) the relative abundance of the lead decay products, and (3) the known half-lives of the uranium isotopes. These numbers, plugged into a formula, yield the radiometric age of the rock.

The authors of the first paper to publish an age for the Catoctin Formation were correct in their commentary that their age determination, using the uranium-lead method, did not make sense, given the particular fossils and the sequence of strata that were present. The error in the age determination lay in the zircon minerals that were analyzed. Not all of them were taken from the Catoctin Formation. Some came from volcanic rocks at Grandfather Mountain in North Carolina, which were mistakenly believed to be correlative with the Catoctin basalts. Also, at the time that the analysis was done (early 1960s), scientists' understanding of radiometric methods was in its infancy; analytical methods have become much more sophisticated since then.

Another method of dating rocks is by the rubidium (Rb)–strontium (Sr) method, which makes use of an unstable isotope of Rb called ^{87}Rb. The radioactive decay of ^{87}Rb results in the formation of a new element, an isotope of strontium called ^{87}Sr. Given a quantity of ^{87}Rb in a rock, precisely half of it will undergo radioactive decay to ^{87}Sr over a period of 48.6 billion years, the half-life of ^{87}Rb.

^{87}Sr is not the only strontium isotope, but it is the only one produced by radioactive decay. We also have ^{84}Sr, ^{86}Sr, and ^{88}Sr, all of which are naturally occurring and whose relative abundances can be expressed in terms of a ratio. For example, for every rock sample containing strontium, the ratio of ^{86}Sr/^{88}Sr is fixed at 0.1194. The ratios of ^{87}Sr to the other strontium isotopes, however, will increase with time because ^{87}Rb is consistently decaying to ^{87}Sr . The higher the ratio, the greater the ^{87}Sr content and the older the age of the rock. For example, imagine that two rocks started out at different times with exactly the same amount of Rb and Sr atoms. Time passes. One rock now has a ^{87}Sr/^{86}Sr ratio of 0.7059 while the other has a ratio of 0.7112. The rock with the higher ratio is older by millions of years.

The use of the Rb-Sr method requires measurement of several things. The rock must be decomposed in a chemical laboratory where the amount of Rb and Sr (in parts per million) is measured. Next, a mass spectrometer is used to determine the abundance of Rb and Sr isotopes. This machine is so sensitive that it can separate ^{87}Sr isotopes from those of ^{88}Sr with the use of a very large magnet. This data from the mass spectrometer, along with the half-life of ^{87}Rb, are plugged into a formula to calculate the age of the rock.

If you can get through the chemical symbols and the math hurdle, the procedure is not overly complex. Basically, for a given initial amount of Rb in a rock, the older the rock, the more ^{87}Sr we find in it. Quite simple. Except for one problem, one big problem for the Catoctin Formation.

Rb is a relatively large atom that will not easily fit into the molecular structure of most silicate minerals. When rocks are heated and disturbed, say by a mountain-building event, and when metamorphic fluids (predominantly water) slowly percolate through them, the Rb tends to be dislodged and transported elsewhere. The isotope ^{87}Rb is also removed and, thus, cannot decay to ^{87}Sr. The Rb-Sr method of dating becomes of little use in rocks subjected to metamorphism.

Metabasalts of the Catoctin Formation were metamorphosed and altered during the middle Ordovician period, about 450 to 470 million years ago, by processes associated with the formation of the Appalachian Mountains. Rb was either removed from the rock or moved from one part of the rock to another by the metamorphic fluids. In the course of my studies, I used a mass spectrometer to measure the Rb and Sr isotope content in dozens of Catoctin samples (I spent two years doing this!). In most samples the Rb and, hence, the Sr concentrations, had been disturbed. Indeed, another geologist, using the Rb-Sr method on other Catoctin samples obtained an age of 420 million years, with a range of uncertainty of plus or minus 70 million years. Realizing that Cambrian-aged fossils occur in rocks overlying the Catoctin Formation, he added the 70 million uncertainty range to the 420 million years and reported a "volcanic" age of 490 million years. Clearly his samples had been affected by metamorphism. ^{87}Rb had been lost from the rocks, so this date was much too young. It was closer to a metamorphic age than to the age of extrusion and crystallization of the volcanic lavas.

Despite the metamorphic alteration and the loss of ^{87}Rb from most Catoctin basalts, some samples from the interior of the lava flows still contain the original igneous minerals and igneous textures. These samples were less altered by metamorphic fluids than samples from the margins of the flows where fluid alteration was intense. Using five of these less-altered samples, plus a few grains of the igneous mineral pyroxene that I extracted from another sample, I obtained some meaningful numbers that yielded an age of 570 million years (Badger and Sinha, 1988). This was the first radiometrically determined age that was in agreement with the stratigraphic and fossil evidence. Since then, geologists from the U.S. Geological Survey (Aleinikoff and others, 1991) have obtained an age of 565 million years from volcanic rocks at

South Mountain, Pennsylvania, that are interpreted to be part of the Catoctin Formation. These two age calculations are in excellent agreement.

One more bit of data helped to confirm our age estimate. Just as I was finishing my Rb and Sr isotope study of the Catoctin metabasalts, two fellow students at Virginia Polytechnic Institute, Ed Simpson and Fred Sundberg (1987), found two fossils in sedimentary rocks overlying the Catoctin Formation in central Virginia. The trace fossil *Rusophycus* was found in the Unicoi Formation, which underlies the Hampton Formation in central Virginia, and the shelly fossil *hyolithid* was found in the Hampton Formation. Both of these fossils are from the early Cambrian time period.

The fossil evidence, the stratigraphic evidence, and the radiometric ages lead to the same conclusion. The volcanic activity that produced the basaltic lavas of the Catoctin Formation in Shenandoah National Park took place during the late Precambrian time period, approximately 570 to 565 million years ago. ■

Glossary

acid-neutralizing capacity

The ability of soil or water to neutralize acids found in precipitation.

albite

A colorless or milky white mineral of the feldspar group containing sodium, with lesser amounts of calcium and potassium.

amygdules

Small cavities or bubbles in volcanic rocks filled with minerals such as calcite, quartz, and feldspar. A rock containing amygdules is referred to as amygdaloidal.

argillaceous

Refers to a sedimentary rock containing a high percentage of clay.

basalt

A dark-colored extrusive igneous rock composed primarily of calcium feldspar (plagioclase), pyroxene, and volcanic glass.

bedding

The layering of sedimentary rocks in beds of varying thickness and character.

bedrock

The general term for the solid rock that underlies soil or other unconsolidated material.

biotite

A black-colored mica commonly found in metamorphic rocks.

breccia

A coarse-grained rock composed of angular broken rock fragments held together by a mineral cement or in a fine-grained matrix.

Cambrian

The geologic time period from about 544 to 500 million years ago. Older time scales put the beginning of the Cambrian at 570 million years, but new data support the more recent date.

chlorite

A dark greenish-colored mica.

columnar jointing

Parallel, prismatic columns, polygonal in cross section, formed as a result of contraction during cooling of basaltic magmas.

conglomerate

A coarse-grained sedimentary rock composed of rounded fragments larger than 2 mm in diameter and set in a fine-grained matrix of sand or silt.

crossbedding

A layering of sedimentary rocks in which the layers are inclined at an angle to the main planes of stratification. This frequently occurs within sand dunes.

dike

A tabular igneous intrusion that cuts across the bedding or foliation of the rock through which it passes.

dip

The angle at which a rock surface, such as bedding or metamorphic layering, is inclined to the horizontal.

epidote

A pistachio- to olive-green, metamorphic mineral containing calcium, iron, aluminum, and silica.

feldspar

A group of abundant rock-forming minerals containing aluminum, silica, and varying amounts of sodium (albite), potassium (orthoclase), and calcium (plagioclase).

float

Loose fragments of rock on the surface of the ground that are no longer attached to the bedrock.

flood basalt

Basaltic lavas that occur as vast accumulations of horizontal or sub-horizontal flows, which erupted in rapid succession over great areas, and have at times flooded sectors of the Earth's surface on a regional scale.

formation

A mappable unit of rock that can be distinguished by its rock type from surrounding rock units.

gneiss

A foliated rock formed by high-grade regional metamorphism, containing bands of minerals (but rarely micas) that give the rock a layered appearance.

gneissic texture

A layering in coarse-grained metamorphic rocks caused by alignment of minerals into thin, irregular bands.

granite

An intrusive igneous rock containing quartz and feldspar, with orthoclase feldspar in greater or approximately equal abundance to plagioclase feldspar.

granodiorite

An intrusive igneous rock containing quartz and feldspar, with plagioclase feldspar in greater abundance than orthoclase feldspar.

greenstone

A basalt that has been altered by metamorphism, commonly containing chlorite, epidote, and albite.

Grenville Mountains

A chain of mountains formed about 1 billion years ago along what is now the eastern part of North America.

Grenville Orogeny

A mountain-building event that affected the eastern side of what is now North America approximately 1 billion years ago.

hematite

An iron oxide, Fe_2O_3, that has a deep reddish color.

hornblende

A black-colored mineral of the amphibole group containing calcium, iron, magnesium, aluminum, and silica.

igneous

A rock that has solidified from molten material.

isotope

An atom containing an abnormal number of neutrons in the nucleus.

laterite

A highly weathered red soil rich in oxides of iron and aluminum.

limestone

A sedimentary rock containing predominantly calcium carbonate.

magnetite

An oxide of iron, Fe_3O_4, that is magnetic.

metabasalt

A basalt that has been altered by metamorphism, commonly containing chlorite, epidote, and albite.

metamorphic

A rock that has been altered by heat, pressure, and hot fluids, usually water, that slowly percolated through the rock.

metamorphism

Mineralogic, chemical, and structural alteration of rock by heat, pressure, and hot fluids at some depth below the surface of the Earth.

micaceous

A rock containing appreciable amounts of mica.

Ordovician

The geologic time period from about 500 to 435 million years ago.

orthoclase feldspar

A colorless, milky white, or sometimes pinkish mineral of the feldspar group containing potassium, with lesser amounts of sodium and calcium.

outcrop

An exposure of bedrock on the surface of the earth.

Pangea

A supercontinent that existed from about 300 to 200 million years ago and included most of the continental crust of the Earth. The present continents were formed when this supercontinent slowly broke apart.

pH

A measure of the acidity or basicity of a solution; a pH of 7 is neutral, below 7 is acidic, and above 7 is basic.

phyllite

A metamorphic rock intermediate in grade between slate and mica schist. Minute crystals of mica give the rock a silky sheen on its surface.

physiographic province

A region of the Earth in which all parts are similar in geologic structure and in which the pattern of landforms differs significantly from that of adjacent regions.

plagioclase feldspar

A colorless, milky white, or sometimes gray mineral of the feldspar group containing calcium, with lesser amounts of sodium and potassium.

plate tectonics

A theory in which the Earth's crust is divided into a number of plates. These plates are in constant motion. Where two plates are colliding with one another, mountains are often formed; where two plates are pulling apart, valleys and ocean basins are formed.

Precambrian

All geologic time from the beginning of the Earth, about 4.6 billion years ago, to the beginning of the Cambrian time period, about 544 million years ago.

pyroxene

A group of dark-colored silicate minerals containing calcium, iron, and magnesium.

quartz

Crystalline silica, SiO_2, the second most common rock-forming mineral (after feldspar).

quartzite

A rock, consisting almost exclusively of quartz, formed by dense compaction of sandstone, usually associated with low-grade metamorphism.

radiometric

A method of dating rocks by measuring the relative abundance of radioactive elements and their daughter products in a rock.

sandstone

A medium-grained sedimentary rock composed of abundant rounded fragments of sand set in a finer-grained matrix (silt or clay) and more or less firmly united by a cementing material, commonly silica, iron oxide, or calcium carbonate.

scoria

Volcanic rock containing vesicles or small air pockets that were formed by gases bubbling out of molten lava.

sedimentary

Rock formed by deposition of fragmental material that originated from weathering of other rocks. This material is usually transported or deposited by water and accumulates in a loose, unconsolidated form that later becomes compacted by overlying material.

shale

A fine-grained sedimentary rock composed of consolidated particles of mud, silt, and clay. The rock breaks readily into thin layers.

silicate melt

Magma containing significant amounts of SiO_2.

silicate mineral

A compound with a crystal structure containing SiO_4 tetrahedra, either isolated or joined through one or more of the oxygen atoms with metallic elements.

siliciclastic

Pertaining to noncarbonate sedimentary rocks which are predominantly composed of silica, such as sandstone, phyllite, or conglomerate.

Skolithos

Burrowing worm from the early Cambrian time period that formed vertical burrows in unconsolidated sand.

slickenlines

Striations on a rock surface formed by friction during movement; commonly found on fault surfaces. A surface containing numerous slickenlines is called a slickenside.

talus slope

A steep slope formed by an accumulation of loose rock fragments, commonly at the base of a cliff.

tectonics

A branch of geology dealing with the broad architecture of the outer part of the Earth. This includes the study of continental and oceanic plates, their movement and interaction, and the study of the mountains formed when two plates converge.

thrust fault

A shallow dipping fault in which one segment of the Earth's crust is pushed up and over another segment of crust.

vesicle

A small cavity in a volcanic rock formed by the entrapment of gas bubbles during solidification of the lava.

vesicular

A volcanic rock containing numerous vesicles.

volcaniclastic

A rock composed of broken fragments of pre-existing volcanic rocks that have been transported from their place of origin by wind or water.

zircon

A mineral, $ZrSiO_4$, that is a common accessory mineral in siliceous volcanic rocks.

References

Aleinikoff, J.N., Zartman, R.E., Rankin, D.W., Lyttle, P.T., Burton, W.C., and McDowell, R.C., 1991, New U-Pb ages for rhyolite of the Catoctin and Mount Rogers Formations - more evidence for two pulses of Iapetan rifting in the Central and Southern Appalachians: *Geol. Soc. of America Abst. with Programs*, v. 23, no. 1, p. 2.

Badger, R. L. and Sinha, A.K., 1988, Age and Sr isotopic signature of the Catoctin volcanic province: implications for subcrustal mantle evolution: *Geology*, v. 16, p. 692-695.

Bates, R.L., and Jackson, J.A., editors, 1980, *Glossary of Geology, 2nd edition*: American Geological Institute, Virginia.

Gathright, T.M., II, 1976, Geology of the Shenandoah National Park, Virginia: *Virginia Division of Mineral Resources Bulletin* 86, 93 p.

Rankin, D.W., 1975, The continental margin of eastern North America in the southern Appalachians: the opening and closing of the proto-Atlantic ocean: *American Journal of Science*, v. 278-A, p. 1-40.

Reed, J.C., Jr., 1955, Catoctin Formation near Luray, Virginia: *Geological Society of America Bulletin*, v. 66, p. 871-896.

Reed, J.C., Jr., and Morgan, Ben, 1971, Chemical alteration and spilitization of the Catoctin greenstones, Shenandoah National Park, Virginia: *Journal of Geology*, v. 79, no. 5, p. 526-548.

Simpson, E.L., and Sundberg, F.A., 1987, Early Cambrian age for synrift deposits of the Chilhowee Group of southwestern Virginia: *Geology*, v. 15, p. 123-126.

Webb, J.R., Deviney, F.A., Galloway, J.N., Rinehart, C.A., Thompson, P.A., and Wilson, S., 1994, The acid-base status of native brook trout streams in the mountains of Virginia: a regional assessment based on the Virginia Trout Stream Sensitivity Study: Dept. of Env. Sciences, Univ. of Virginia.

Webb, J.R., Deviney, F.A., and Galloway, J.N., in review, Integrated data assessment project, part B, stream-water composition component: Dept. of Env. Sciences, Univ. of Virginia.

Index

Page numbers in **bold** refer to maps.
Page numbers in *italics* refer to photos.

Leading the way

FALCONGUIDES are available for where-to-go hiking, mountain biking, rock climbing, walking, scenic driving, fishing, rockhounding, paddling, birding, wildlife viewing, and camping. The following titles are currently available, but this list grows every year. We also have FalconGuides on essential outdoor skills and subjects and field identification. For a free catalog with a complete list of titles, call FALCON toll-free at 1-800-582-2665.

Rockhounding Guides

Rockhounding Arizona
Rockhounding California
Rockhounding Colorado
Rockhounding Montana
Rockhounding Nevada
Rockhounding New Mexico
Rockhounding Texas
Rockhounding Utah
Rockhounding Wyoming

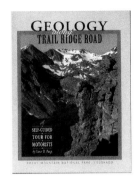

Also Available:

Geology along Trail Ridge Road:
a Self-Guided Tour for Motorists

Other regional FALCONGUIDES

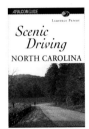

Hiking Guides

Best Easy Day Hikes Shenandoah
Hiking Southern New England
Hiking Maine
Hiking New Hampshire
Hiking New York
Hiking Pennsylvania
Hiking Utah
Hiking Florida
Hiking Georgia
Hiking North Carolina
Hiking South Carolina
Hiking Tennessee
Hiking Shenandoah National Park
Hiking Virginia

Scenic Driving Guides

Scenic Driving New England
Scenic Driving Florida
Scenic Driving Georgia
Scenic Driving North Carolina

On My Mind Series

Florida on My Mind
Georgia on My Mind
North Carolina on My Mind
Virginia on My Mind

The Insiders' Guide to:

Civil War Sites
    in the Eastern Theater
Richmond
Virginia's Blue Ridge
Virginia's Chesapeake Bay
Washington, D.C.

Wildlife Viewing Guides

Florida   Pennsylvania
Massachusetts   Tennessee
New Hampshire   Vermont
New Jersey   Virginia
New York
North Carolina

FALCON®

*To order any of these books, check with your local bookseller
or call FALCON® at 1-800-582-2665.
Visit us on the world wide web at:
www.falconguide.com*